The Universe of *Escherichia coli*

Edited by Marjanca Starčič Erjavec

Published in London, United Kingdom

IntechOpen

Supporting open minds since 2005

The Universe of *Escherichia coli*
http://dx.doi.org/10.5772/intechopen.73717
Edited by Marjanca Starčič Erjavec

Contributors
Elif Bozcal, Isabel Scaletsky, Wan Keung Raymond Wong, Masahide Yasuda, Jin Matsumoto, Tomoko Matsumoto, Kazuya Yasuda, Angel Adrian Cataldi, Mariano Larzabal, Gozde Ekici, Emek Dumen, Marjanca Starčič Erjavec

First published in London, United Kingdom, 2019 by IntechOpen
IntechOpen is the global imprint of INTECHOPEN LIMITED, registered in England and Wales, registration number: 11086078, The Shard, 25th floor, 32 London Bridge Street
London, SE19SG – United Kingdom
Printed in Croatia

British Library Cataloguing-in-Publication Data
A catalogue record for this book is available from the British Library

Additional hard and PDF copies can be obtained from orders@intechopen.com

The Universe of *Escherichia coli*
Edited by Marjanca Starčič Erjavec
p. cm.
Print ISBN 978-1-83881-152-5
Online ISBN 978-1-83881-153-2
eBook (PDF) ISBN 978-1-83881-154-9

For EU product safety concerns:
IN TECH d.o.o., Prolaz Marije Krucifikse Kozulić 3, 51000 Rijeka, Croatia, info@intechopen.com or visit our website at intechopen.com.

Meet the editor

Marjanca Starčič Erjavec, PhD, is a professor of molecular biology at the Department of Biology, Biotechnical Faculty, University of Ljubljana, Slovenia. She finished the BSc study of biology and the MSc study of biochemistry and molecular biology at the University of Ljubljana, Slovenia. She defended her PhD thesis in the field of bacterial molecular genetics at the Utrecht University, The Netherlands. She is involved in teaching several molecular biological and genetic courses. She was a visiting professor at several universities in Europe and USA and also at the Institute of Ecology and Genetics of Microorganisms, Ural Branch Russian Academy of Science, Perm, Russia. She conducts research on horizontal gene transfer, including plasmids, natural and clinical *E. coli* strains, and microbiota.

Contents

Preface

The title of the book "The Universe of *Escherichia coli*" aims to present and emphasize the huge diversity of this bacterial species and our efforts to prevent the infections caused by *E. coli*. *E. coli* is on one hand a well-known commensal species, as it is part of the gut microbiota of humans and other warm blooded organisms. Probiotic strains of *E. coli* do exist and they are successfully used for improving host's health. Also many "workhorse" *E. coli* strain exist that are employed in laboratory and biotechnology settings. But on the other hand, *E. coli* is on the list of biological agents – it is grouped in the human pathogen hazard group 2 and some *E. coli* strains (enterohemorrhagic strains) even in the hazard group 3. Furthermore, *E. coli* belongs to the bacterial pathogens with the broadest disease spectrum, as it can cause intestinal and also extraintestinal infections at many anatomical sites. And *E. coli* is also known to be an important animal pathogen. Therefore many efforts are undertaken to prevent *E. coli* infections, among them food safety, vaccines, but also new alternative antimicrobial agents are searched for.

Hence the book has two sections, one dealing with the versatility of *E. coli* and the second dealing with our efforts to prevent *E. coli* infections. For both sections chapters were selected to represent the main topics.

Laboratory *E. coli* strains are categorized as non-pathogenic strains that can be safely used. So *E. coli* is a well-known and much used laboratory "workhorse", used as a model organism in many basic research areas. But *E. coli* is used also in biotechnology, where it is due to its rapid growth and easy handling, a much appreciated host organism, so many recombinant proteins are made by *E. coli* (see Chapter 1). Pathogenic *E. coli* strains are very diverse and associated with many types of infections in humans and animals. Among them a very prominent group are the enteropathogenic *E. coli*, which have been associated with outbreaks of diarrhoea (see Chapter 2). The versatility of *E. coli* is presented also by a large number of different mobile genetic elements such as transposons, plasmids, and insertion sequences that contribute to the plasticity of *E. coli* genome (see Chapter 3).

As *E. coli* is a well-known pathogen that becomes more and more resistant to antibiotics, we are already faced with difficulties in treatment of some *E. coli* infections, namely infections caused by the extended-spectrum beta-lactamases producing and carbapenem resistant *E. coli*. In this light, preventing *E. coli* infections by different means gains on importance. As humans can be infected by ingestion and/or not proper handling of with pathogenic *E. coli* contaminated food, special care has to be devoted to food safety and hence there is a chapter in the book dealing with this topic (see Chapter 4). As vaccines are a mean of possible protection against pathogenic *E. coli* that is not depending on the antimicrobial resistance pattern, a special chapter in the book is dealing with human and veterinary vaccines against pathogenic *E. coli* (see Chapter 7). WHO has put *Enterobacteriaceae* and among them *E. coli* on the list of bacteria for which new antimicrobial agents are urgently needed and grouped them into the Priority 1: CRITICAL group. So, in the last chapter the possibilities of photodynamic inactivation of *E. coli* with cationic porphyrin sensitizers are presented (see Chapter 8).

To conclude, I do hope that these selected chapters, even though small in number, do represent the extent of *E. coli* universe and my sincere thanks go to all the authors from many countries, which contributed these chapters. I also want to express my thanks to InTech, who gave the opportunity of publishing this book and to all the staff at InTech who helped in the process of making this book seeing the light.

Prof. dr. Marjanca Starčič Erjavec
Department of Biology, Biotechnical Faculty,
University of Ljubljana,
Slovenia

Section 1

Introduction

Chapter 1

Introductory Chapter: The Versatile *Escherichia coli*

Marjanca Starčič Erjavec

1. Introduction

There are not so many organisms that are so well studied and researched as the bacterium *Escherichia coli* (*E. coli*). Since its discovery in 1885, it was used in research, and by end of 2018, there are now already 368,071 publications in PubMed about *E. coli* [1]. **Figure 1**, presenting data about number of publications found in PubMed for the search term "*Escherichia coli*" in the time frame from 1932 to 2018, clearly demonstrates the high and still growing research interest in this microbe.

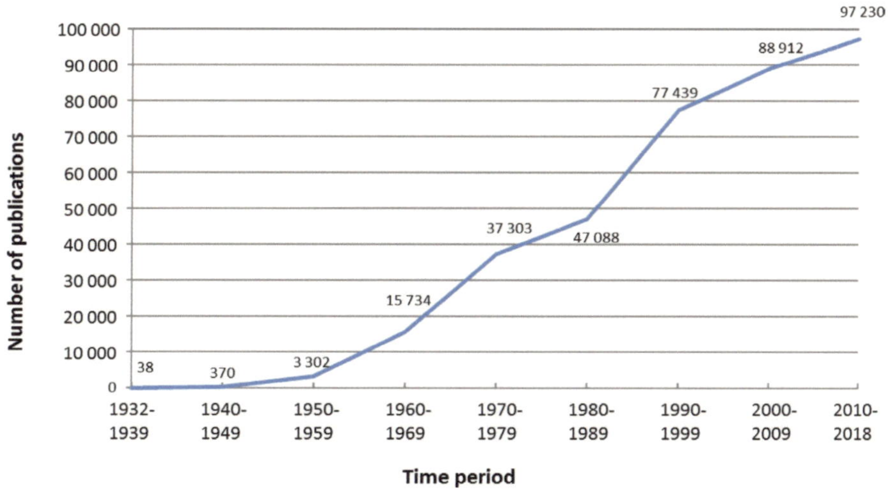

Figure 1.
Number of publications in PubMed for the search term "Escherichia coli" in the time frame from January 01, 1932 to December 31, 2018 [1].

2. The discovery of *Escherichia coli*

The bacterium *E. coli* was discovered by the German-Austrian pediatrician Dr. Theodor Escherich (1857–1911) in 1885 [2]. He conducted examinations of neonate's meconium and feces of breast-fed infants with the aim to gain insight into the development of intestinal "flora." In preparations of meconium and stool samples under the microscope, he observed "slender short rods" of the size of 1–5 μm in length and 0.3–0.4 μm in width, which he named *Bacterium coli commune* (**Figure 2**). Further, he cultured these bacteria on agar and blood serum plates, where these bacteria grew as white, non-liquefying colonies. He also showed that these bacteria slowly cause milk to be clotted, as a result of acid formation, and

IntechOpen

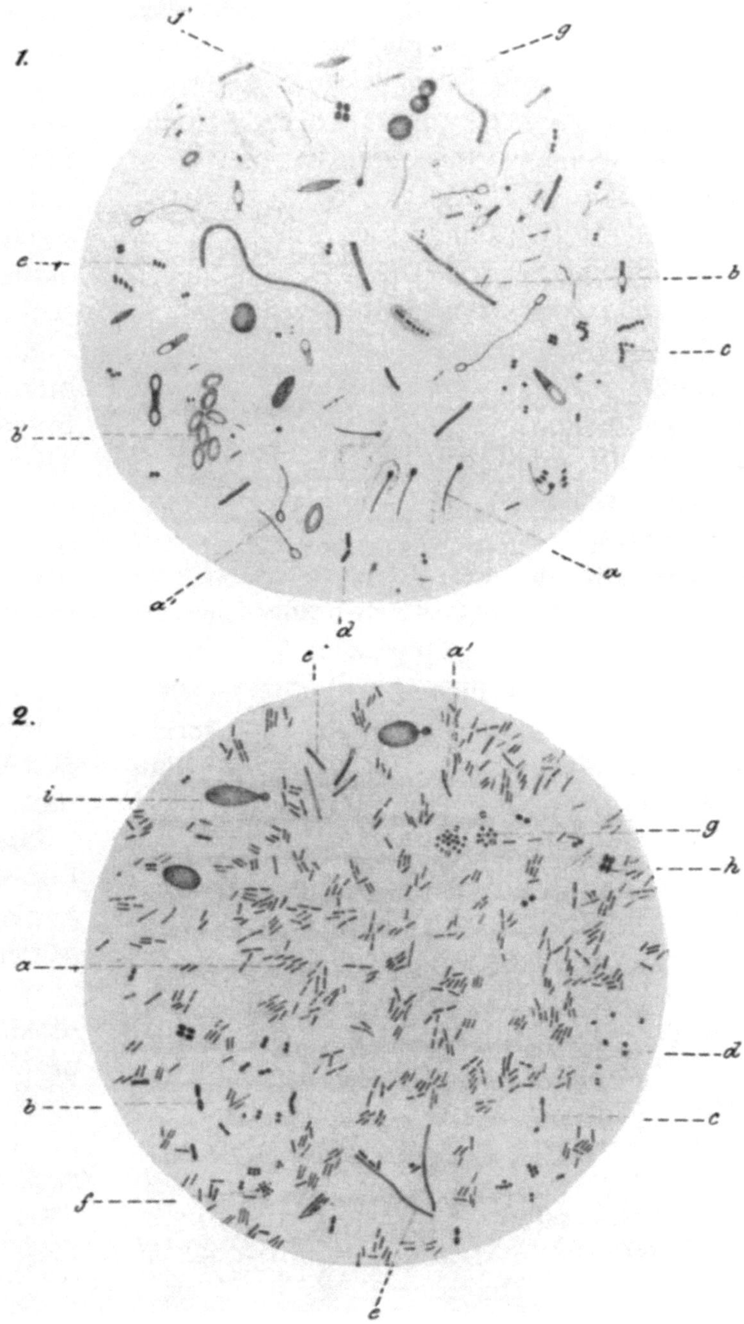

Figure 2.
Escherich's drawing of the stool bacteria, as seen under light microscope [4]. Panel 1: Preparation of a meconium of a 27-hour-old infant. The E. coli as Bacterium coli commune is represented under d. Panel 2: Preparation of a stool of a 2-month-old healthy breast-fed child. The E. coli as Bacterium coli commune is represented under a and a'.

demonstrated that these bacteria have fermentative ability. He also performed the Gram method of staining and revealed that these bacteria rapidly take color with all aniline dyes but lose the color after treatment with potassium iodide and alcohol [2]. Later, in 1919, the bacterium was renamed after its discoverer by Castellani and Chalmers and became *Escherichia coli* [3].

3. Characteristics of *Escherichia coli*

3.1 Basic characteristics

The bacterium *E. coli* (**Figure 3**) belongs into the family of *Enterobacteriaceae*. It is a Gram-negative rod-shaped bacterium, non-sporulating, nonmotile or motile by peritrichous flagella, chemoorganotrophic, facultative anaerobic, producing acid from glucose, catalase positive, oxidase negative, and mesophilic [5].

E. coli is a well-known commensal bacterium that is among the first colonizing bacteria of the gut after birth. It is a highly successful competitor in the human gut and is comprising the most abundant facultative anaerobe of the human intestinal microbiota [7]. As it is a facultative anaerobe, it survives when released to the environment and can be spread to new hosts. *E. coli* is thus an important component of the biosphere [8].

Even though *E. coli* is a well-known commensal bacterium, many pathogenic strains of *E. coli* do exist. Several highly adapted *E. coli* clones have acquired specific virulence factors, which confer an increased ability to adapt to new niches and allow them to cause a broad spectrum of disease, and intestinal and also extraintestinal infections [7].

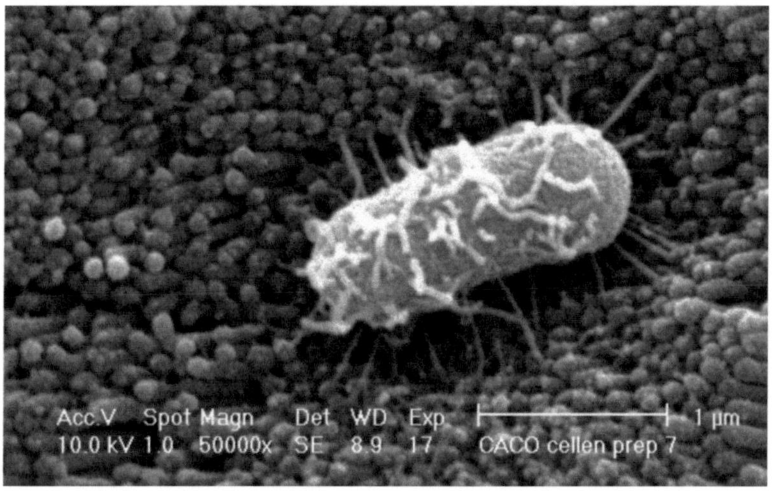

Figure 3.
Scanning electron microscopy of a single bacterial E. coli cell adhering to 19-day-old Caco-2 cells [6].

3.2 The *E. coli* genome

The first complete *E. coli* genome sequence was the sequence of the K-12 MG1655 strain of *E. coli*, published in 1997. The sequenced strain has been maintained as a laboratory strain with minimal genetic manipulation, having only been cured of the temperate bacteriophage lambda and F plasmid. The published genome has 4,639,221 base pairs. Protein-coding genes account for 87.8% of the genome, 0.8% encodes stable RNAs, and 0.7% consists of noncoding repeats. Eleven percent of the genome are involved in regulation of gene expression and also other functions [9]. A circular map of the *E. coli* genome is represented in **Figure 4**.

The map is based on the K-12 MG1655 sequence data as deposited in GenBank (Accession number NC_000913) [10]. The multiplier for the ticks is 1e-6 (1.0 represents 1,000,000). In blue, the forward genes are shown, in purple the reverse genes, tRNA genes in orange, and rRNA genes in red. The map was drawn with

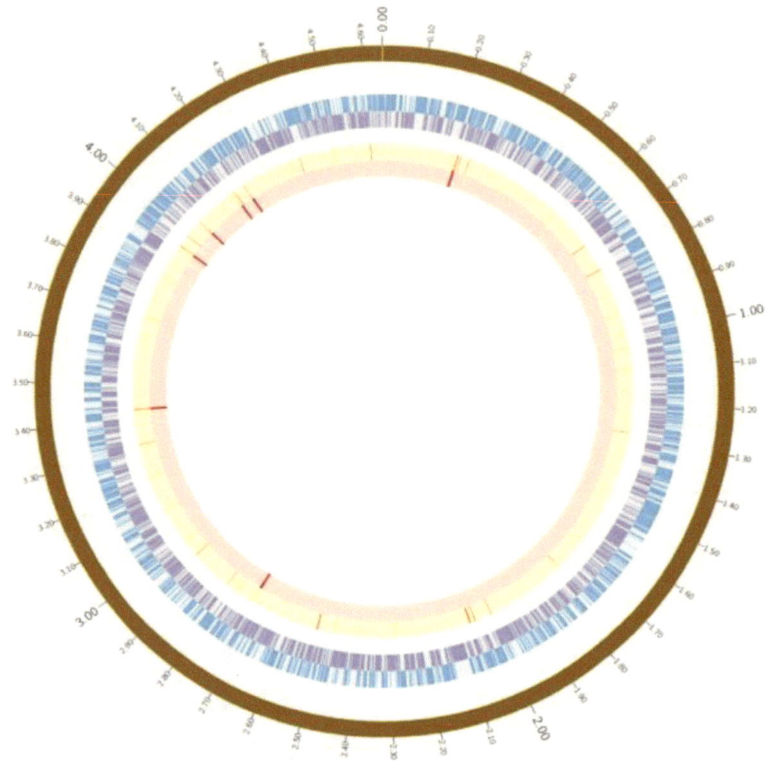

Figure 4.
Circular map of the E. coli K-12 MG1655 strain.

the online tool ClicO FS, available at the Internet site http://www.codoncloud.com:3000/home [11, 12].

Genomes of pathogenic *E. coli* strains are general bigger, as the pathogenic strains need several special properties, so-called virulence factors. These are encoded in the virulence-associated genes (VAGs), which are frequently clustered in DNA regions called pathogenicity islands (PAIs) [13]. Often the pathogenic strains possess also extrachromosomal DNA elements, i.e., plasmids, that can also carry additional VAGs [7]. Some examples of genomes of pathogenic strains in comparison with the K-12 MG1655 strain are given in **Table 1**.

Data in the table are based on data available in the genome database of the National Center for Biotechnology Information (Internet site: www.ncbi.nlm.nih.gov) [14].

The most famous *E. coli* plasmid is the plasmid F (**Figure 5**). It is the paradigm plasmid for plasmid-specified transfer systems, as bacterial conjugation was first identified as a function of the F plasmid. Further, this plasmid was used to develop many of the genetic techniques commonly used to dissect prokaryotic systems, and F product analysis has been central in elucidating the basic mechanisms of plasmid replication and transmission [15].

F plasmid has two functional replication regions, RepFIA and RepFIB. The RepFIA region is believed to be primarily responsible for the typical replication properties of F [16]. The secondary replication region, RepFIB, is independently functional and can perform replication in the absence of RepFIA. F plasmid has also remnants of a third replication region, RepFIC, whose function was abolished by transposition of Tn*1000* into this replication region [17]. Apart from Tn*1000* also insertion sequences IS2 and IS3 are carried on F plasmid [16]. The plasmid-specified transfer system is encoded in the *tra* region, starting with the origin of transfer (*oriT*) [15].

E. coli strain	Associated with infection	Chromosome size (Mbp)	Number of genes in the chromosome	Plasmids	Plasmid size (bp)	Number of genes on the plasmid
K-12 MG1655	/	4.64	4.566	/	/	/
O157:H7 Sakai	Hemorrhagic diarrhea	5.5	5.329	pO157	92.721	85
				pOSAK1	3306	3
O7:K1 IAI39	Urinary tract infection	5.13	5.092	/	/	/
O83:H1 NRG 857C	Crohn's disease	4.75	4.532	pO83_CORR	147.060	154
O104:H4 2011C-3493 ASM29945v1	Hemolytic-uremic syndrome	5.27	5.081	pG-EA11	1549	1
				pAA-EA11	74.217	82
				pESBL-EA11	88.544	94
UMN026	Urinary tract infection	5.2	5.096	p1ESCUM	122.301	156
				p2ESCUM	33.809	49

Table 1.
Genomes of different E. coli strains.

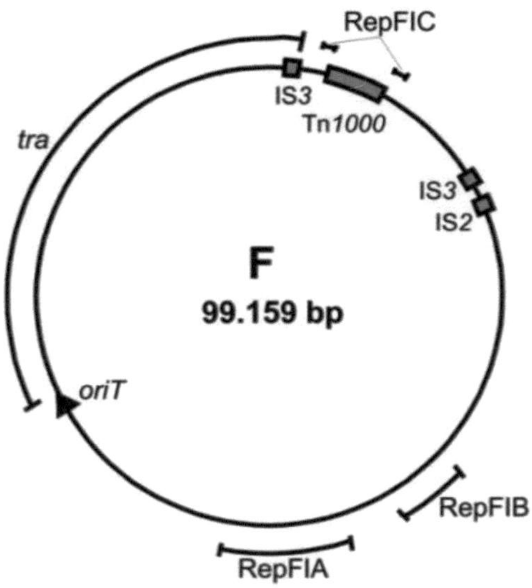

Figure 5.
Map of the E. coli F plasmid. The map was drawn based on the complete nucleotide sequence of the F plasmid as deposited in GenBank [18].

3.3 The phylogenetic groups of *E. coli*

The *E. coli* species has an extensive genetic substructure and the methods to assess the phylogenetic relationship among *E. coli* strains evolved during the time. In the pre-molecular era, the *E. coli* diversity was studied by serotyping. Serotyping studies showed that the somatic (O) antigen, the flagellar (H) antigen, and to a lesser extent the capsular (K) antigen are useful in distinguishing *E. coli* strains [19]. The *E. coli* serotyping is complex—173 O antigens, 80 K antigens, and 56 H antigens are known—and the O, K, and H antigens can be found in nature in many of the possible combinations. The final number of *E. coli* serotypes is therefore very high, 50,000–100,000 or more [20].

The molecular studies of *E. coli* diversity began with the measurement of variations in electrophoretic mobility of enzymes derived from different *E. coli* strains [21]. In 1980s the multi-locus enzyme electrophoresis (MLEE) became the common technique for the study of bacterial diversity. It was found that *E. coli* populations evolve in a clonal manner, with recombination playing a limited role, and it also became clear that genetically distant strains can have the same serotype and that closely related strains may have different serotypes [19]. Based on the MLEE studies of 38 enzyme loci, four major phylogenetic groups among *E. coli* were found: A, B1, B2, and D [22]. Clermont et al. [23] established a method of rapid and simple determination of the *E. coli* phylogenetic groups by a triplex PCR. This genotyping method is based on the amplification of a 279 bp fragment of the *chuA* gene; a 211 bp fragment of the *yjaA* gene; and a 152 bp fragment of TSPE4.C2, a noncoding region of the genome. The presence or absence of combinations of these three amplicons is used to assign the *E. coli* to the phylogenetic groups: A, B1, B2, or D (**Figure 6**).

However, subsequently, on the basis of multi-locus sequence typing and complete genome data, additional *E. coli* phylogenetic groups were recognized [24, 25]. The number of defined phylogenetic groups thus rose to eight (A, B1, B2, C, D, E, F that belongs to *E. coli* sensu stricto, and the eighth—the *Escherichia* cryptic clade I). Clermont et al. [26] thus revised their method to encompass the newly described phylogenetic groups. To enable identification of the F phylogenetic group, the new extended PCR phylotyping method employs an additional gene target, *arpA*, which serves also as an internal control for DNA quality. Thus, the revised PCR method is based on a quadruplex PCR, and if required, additional single PCR reactions are employed to distinguish between E and clade I, A or C, and D or E phylo-group [26] (**Figure 7**).

Two collections of human fecal isolates were screened using the quadruplex phylo-group assignment method demonstrating that 12.8% of *E. coli* isolates belonged to the newly described phylo-groups C, E, F, and clade I and that strains assigned to phylo-groups A and D by the triplex method are worth to be retested by the quadruplex method, as it is likely that they are going to be reclassified [26]. Logue et al. [27] performed a comparative analysis of phylogenetic assignment of human and avian extraintestinal pathogenic (ExPEC) and fecal commensal *E. coli* (FEC) strains and showed that a total 13.05% of studied human *E. coli* strains and 40.49% of avian *E. coli* strains had to be

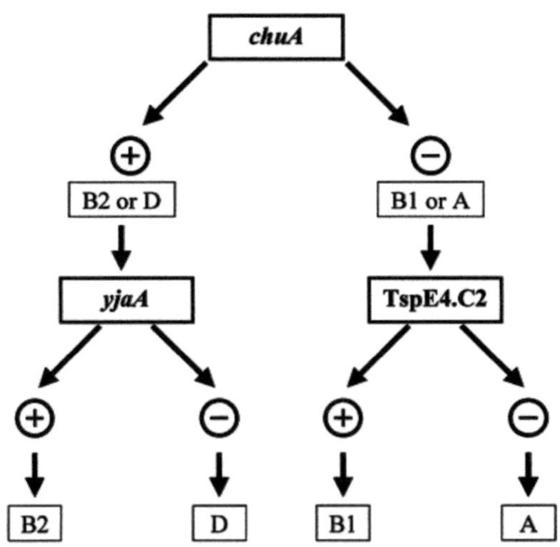

Figure 6.
Dichotomous decision tree to determine the phylogenetic group by the Clermont triplex PCR method [23].

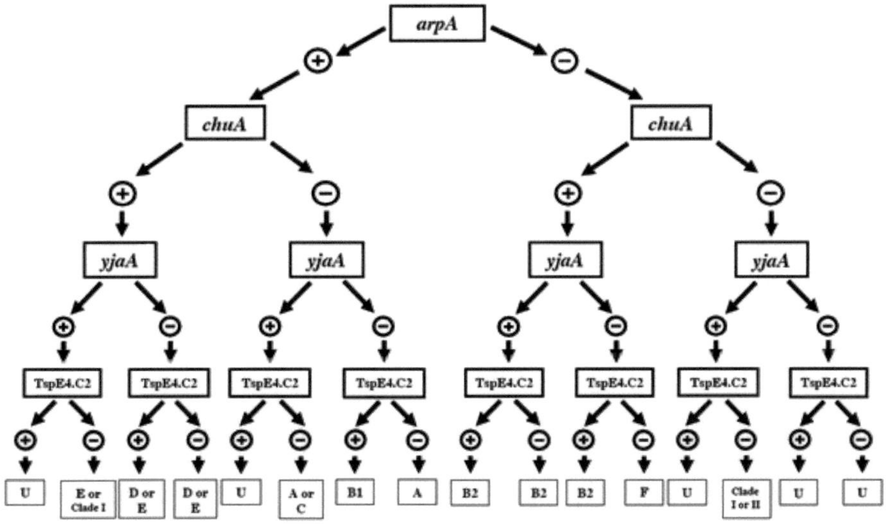

Figure 7.
Dichotomous decision tree to determine the phylogenetic group by the Clermont quadruplex PCR method [26].

reclassified. Another study using human *E. coli* strains isolated from skin and soft-tissue infections and fecal *E. coli* strains from healthy humans and also avian and brown bear fecal strains revealed that 27.60% of human, 23.33% of avian, and 70.93% of brown bear strains had to be reclassified. Moreover, a high number (12.22%) of reclassifications from the previous phylo-groups to the non-typeable (NT) group were observed among the avian fecal strains of this study. Further, a survey performed on other published data by Starčič Erjavec et al. [28] showed that also a number of other studies report occurrence of NT strains by the quadruplex method, for example, a study including 140 uropathogenic *E. coli* strains from Iran reported 27.14% of NT strains [29]. These data emphasizes that there is a need to search for more *E. coli* strains from novel environments (new hosts in not yet explored geographic regions) and to revise the PCR phylotyping method again in order to type these NT strains.

3.4 The commensal *E. coli*

As *E. coli* is a facultative anaerobe, and among the first gut colonizers, these bacteria help to establish the anaerobic environment of the gut that enables the further colonization of the gut by anaerobic bacteria [30]. After the *E. coli* colonization, usually the host and *E. coli* coexist in mutual benefit for decades [7]. *E. coli* gets "food and shelter," and the host benefits due to the *E. coli* vitamin K production and the so-called colonization resistance. Colonization resistance is the phenomenon of protection against colonization by pathogenic bacteria, including pathogenic *E. coli* [31]. The niche of the commensal *E. coli* is the mucous layer of the colon [7]. On average five different commensal *E. coli* strains colonize a human host at any given time [32]. As host and the *E. coli* profits from their association, these *E. coli* could be also designated as mutualistic *E. coli*.

3.5 The pathogenic *E. coli*

E. coli is also a medically important species, as it is involved in many different types of infections. Two major groups of pathogenic *E. coli* exists: the intestinal pathogenic *E. coli* (IPEC), associated with infections of the gastrointestinal tract, and the extraintestinal pathogenic *E. coli* (ExPEC), associated with infections of extraintestinal anatomic sites [7]. The medical diversity of this species is nicely exhibited by its classification of pathogenic *E. coli* (**Figure 8**), the so-called *E. coli* pathotypes.

The versatility of pathogenic *E. coli* strains depends on their genetic makeup, on the presence of so-called virulence genes, and possession of such genes distinguishes pathogenic from nonpathogenic bacteria [34]. Virulence factors help bacteria to (1) invade the host, (2) cause disease, and (3) evade host defenses [35].

3.5.1 Adhesins and invasins

Once a bacterium reaches the host surface, in order to colonize, it must adhere to host cells. For this purpose bacteria have different fimbrial and afimbrial adhesins. Fimbrial adhesins are rod-shaped protein structures, which consists primarily of an ordered array of single protein subunits, which build a long cylindrical structure. At the top, there are proteins, adhesins, which mediate the adherence to the host's molecules. A fimbrial adhesin is thus a structure that extends outward from the bacterial surface and establishes the contact between the bacterial surface and the surface of the host cells. Afimbrial adhesins are surface proteins important for tighter binding of bacteria to host cells. Some bacteria have evolved mechanisms for entering nonphagocytic host cells. Bacterial surface proteins that provoke actin rearrangements and thereby incite the phagocytic ingestion of the bacterium by host cells are called invasins [36]. The most known *E. coli* adhesins and invasins are presented in **Table 2**.

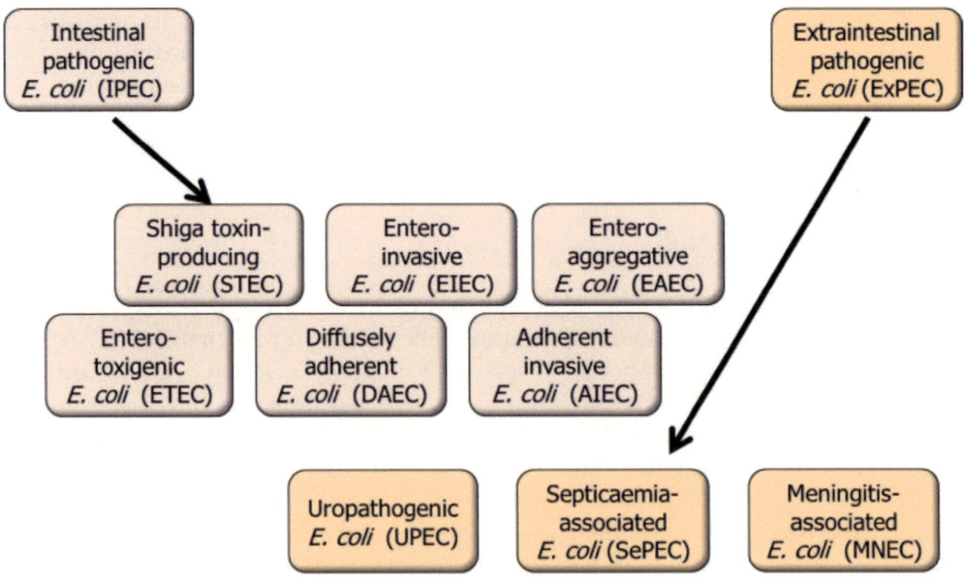

Figure 8.
Classification of pathogenic E. coli, based on Roy et al. [33]. The IPEC are also designated as diarrheagenic E. coli (DEC)—Although not all of the subtypes in this group necessarily cause diarrhea. STEC that cause hemorrhagic colitis and/or the hemolytic uremic syndrome are called EHEC—For enterohemorrhagic E. coli. Among ExPEC also strains associated with pneumonia, skin and soft-tissues, and infections of many other extraintestinal anatomic sites are present, though they are not yet established as separate pathotypes.

Adhesin/invasin	Most commonly tested virulence (associated) genes
Type 1 fimbriae (Fim)	*fimH*
P fimbriae (Pap/Prf)	*papC*, *papG*
S/F1C fimbriae (Sfa/Foc)	*sfa/focDE*
N-Acetyl-D-glucosamine-specific fimbriae (Gaf)	*gafD*
M-Agglutinin (Bma)	*bmaE*
Bifunctional enterobactin receptor/adhesin (Iha)	*iha*
Afimbrial adhesin (Afa)	*afa/draBC*
Invasion of brain endothelium (IbeA)	*ibeA*
Colonization factor antigen I (CFA/I)	*cfaB*
Bundle-forming pili (BFP)	*bfpA*
Intimin	*eaeA*
Aggregative adherence fimbriae (AAF/I)	*aaf/I*

Table 2.
Typical adhesins and invasins of pathogenic E. coli strains.

3.5.2 Iron acquisition mechanisms

Iron is essential for bacterial growth, but iron concentrations in nature are generally quite low, particularly low in host organism. To survive in the host organism, bacteria must have some mechanisms for acquiring iron. The best studied type of bacterial iron acquisition is the siderophores. These are low-molecular-weight compounds that chelate iron with very high affinity [36]. The most known *E. coli* iron uptake systems are presented in **Table 3**.

3.5.3 Systems to evade host immune response

The healthy host usually has multilayered defenses that prevent the establishment of bacterial infection. Among the most effective of these defenses is the immune response. However, bacteria have evolved systems to avoid, subvert, or circumvent innate host defenses and to evade acquired specific immune responses of the host [34]. A capsule is a loose, relatively unstructured network of polymers that covers the surface of a bacterium. The role of capsules in bacterial virulence is to protect bacteria from the host's inflammatory response [36]. Further, increased serum resistance is often found among pathogenic bacteria, especially those associated with systemic infections [36]. Serum resistance is the ability to prevent complement activation on the bacterial cell surface and to inhibit insertion of the membrane attack complex into the bacterial membrane [34]. The feature is often based on the modifications in lipopolysaccharide (LPS), which can be of two types: either attachment of sialic acid to LPS O antigen or changes in the LPS O antigen side chain [36]. However, other proteins can also be implicated in increased serum resistance; for example, the TraT protein of the surface exclusion complex involved in conjugation [37]. Another important protein of pathogenic *E. coli* is the Toll/interleukin-1 receptor domain-containing protein (Tcp) that interferes with the TLR signaling system of the innate immunity [38]. The most known *E. coli* systems to evade host immune response are presented in **Table 4**.

Iron uptake system	Most commonly tested virulence (associated) genes
Aerobactin (Iuc)	*iucD, iutA*
Yersiniabactin (Ybt)	*fyuA, irp2*
Salmochelin (Iro)	*iroCD, iroN*
Siderophore receptor IreA	*ireA*
Temperature sensitive hemagglutinin (Tsh)—in birds, Hemoglobin protease (Hbp)—in humans	*tsh, hbp*
Periplasmic iron binding protein (SitA)	*sitA*
Ferrichrome-iron receptor (Fhu)	*fhuA*

Table 3.
Typical iron uptake systems of pathogenic E. coli strains.

Host immunity evading system	Most commonly tested virulence (associated) genes
Group II capsule including K1 and K5 capsules	*kpsMT II*
Conjugal transfer surface exclusion protein (TraT)	*traT*
Outer membrane protease T (OmpT)	*ompT, APEC-ompT*
Increased serum survival (Iss)	*iss*
Suppression of innate immunity (Toll/interleukin-1 receptor domain-containing protein Tcp)	*tcpC*

Table 4.
Typical host immunity evading systems of pathogenic E. coli strains.

3.5.4 Toxins

Toxins are the virulence factors that damage the host. Exotoxins are toxic bacterial proteins that are excreted into the medium by growing bacteria or localized in the bacterial cytoplasm or periplasm and released during bacterial lysis. Exotoxins vary considerably in their activities and the target host cell types [36]. The most known *E. coli* toxins (exotoxins) are presented in **Table 5**.

Toxins	Most commonly tested virulence (associated) genes
alpha-Hemolysin (HlyA)	*hlyA*
Cytotoxic necrotizing factor 1 (CNF-1)	*cnf1*
Cytolethal distending toxin IV (CDT 1)	*cdtB*
Uropathogenic specific protein (Usp)	*usp*
Colibactin (Clb)	*clbAQ*
Serine protease autotransporters Sat, Pic	*sat, picU*
Heat-stable toxins (STa, STb)	*stIa/stIb*
Heat-labile toxin I (LTI), heat-labile toxin II (LTII)	*eltI, eltIIa*
Shiga toxin 1 (Stx1), Shiga toxin 2 (Stx2)	*stxI, stxII*
EHEC hemolysin (Ehx)	*ehxA*
Low-MW heat-stable toxin (EAST1)	*astA*

Table 5.
Typical toxins (exotoxins) of pathogenic E. coli strains.

However, *E. coli* possess also an endotoxin, namely, the lipopolysaccharide, which is an integral component of the outer membrane of Gram-negative bacteria. The lipid portion (lipid A) is embedded in the outer membrane, with the core and O antigen portions extending outward from the bacterial surface. Lipid A is the toxic portion of the molecule, and it exerts its effects only when bacteria are lysed. The toxicity of lipid A resides primarily in its ability to activate, complement, and stimulate the release of bioactive host proteins, such as cytokines [36].

3.6 The antibiotic-resistant *E. coli*

Antibiotics are low-molecular-weight compounds that kill or inhibit growth of bacteria [36]. Antibiotic treatment is one of the main approaches of modern medicine to combat bacterial infections, including also *E. coli* infections [39]. However, bacteria evolved different mechanisms that confer resistances to antibiotics. Resistant bacteria are able to either (i) modify/degrade the antibiotic, (ii) actively transport the antibiotic out of the cell or prevent its intake, (iii) sequester the antibiotic by special proteins, or (iv) modify, bypass, or protect the target [40]. The emergence, spread, and persistence of resistant and even multidrug-resistant (MDR) bacteria or "superbugs", also among *E. coli*, are now posing a serious global health threat of growing concern [39]. The antimicrobial resistance surveillance data of European Centre for Disease Prevention and Control (ECDC) also showed the increase in antibiotic resistance among invasive *E. coli* isolates (**Figure 9**).

The mechanisms of resistance to antibiotics are encoded in resistance genes. A list of typical *E. coli* resistance genes is given in **Table 6**.

As many of the resistance genes are encoded on conjugative plasmids or conjugative transposons, they are easily transferred between different bacteria and hence spread in the population [36].

3.7 The bacteriocinogenic *E. coli*

Bacteriocins are ribosomally synthesized, proteinaceous substances that inhibit the growth of closely related species through numerous mechanisms [51].

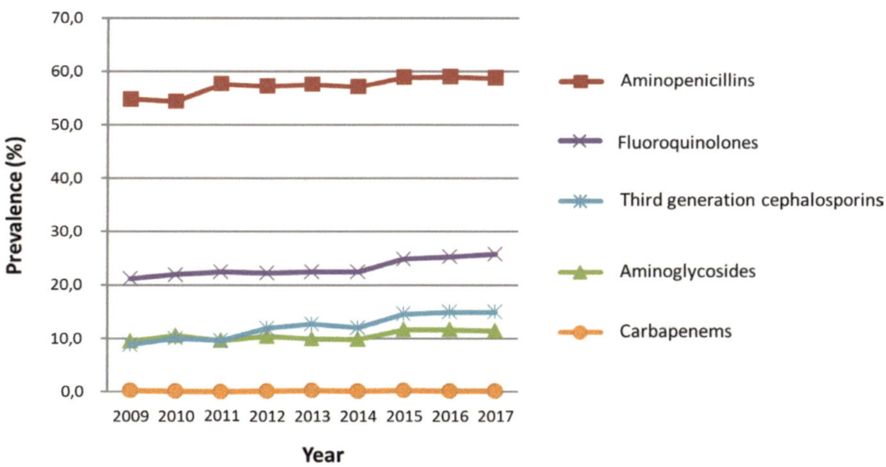

Figure 9.
Prevalence of invasive E. coli isolates with antimicrobial resistance to aminopenicillins, fluoroquinolones, third-generation cephalosporins, aminoglycosides, and carbapenems—the population weighted mean EU/EEA is shown. The prevalence of antimicrobial resistance to carbapenems in 2009 and 2011 was 0%, in 2012 <0.1%, in 2013 and 2015 0.2%, and in 2014, 2016, and 2017 0.1% [41–49].

Resistance gene(s)	Antibiotic class	Resistance to
strA [aph(3')-Ib], *strB [aph(6')-Id]*	Aminoglycosides	STR
aadA1, aadA2, aadA5, aadA7, aadA24	Aminoglycosides	STR
aph(3')-Ia	Aminoglycosides	KAN
aac(3')-VI, aac(3')-IId	Aminoglycosides	GEN
bla_{TEM-1}	β-Lactams	AMP
bla_{OXA-1}	β-Lactams	AMP
bla_{CMY-2}	β-Lactams	AMC, AMP, CRO, FOX, TIO
ampC	β-Lactams	AMC, AMP, FOX
sul1, sul2, sul3	Folate synthesis inhibitors	FIS
dfrA1, dfrA5, dfrA12, dfrA17	Folate synthesis inhibitors	SXT
mphA	Macrolides	AZM
floR	Phenicols	CHL
cmlA	Phenicols	CHL
catA1, catB3	Phenicols	CHL
qnrB2, qnrB6, qnrS2	Quinolones	NAL, CIP
tet(A), *tet*(B), *tet*(C), *tet*(D), *tet*(M)	Tetracyclines	TET

STR, streptomycin; KAN, kanamycin; GEN, gentamicin; AMP, ampicillin; AMC, amoxicillin/clavulanic acid; CRO, ceftriaxone; FOX, cefoxitin; TIO, ceftiofur; FIS, sulfisoxazole; SXT, trimethoprim/sulfamethoxazole; AZM, azithromycin; CHL, chloramphenicol; NAL, nalidixic acid; CIP, ciprofloxacin; TET, tetracycline [50].

Table 6.
Typical E. coli resistance genes.

They are a heterogeneous group of particles with different morphological and biochemical entities. They range from a simple protein to a high molecular weight complex [52]. The bacteriocins with molecular masses below 10 kDa are designated as microcins [53]. Bacteriocins are potent toxins that are usually produced during stressful conditions and result in the rapid elimination of neighboring bacterial cells that are not immune or resistant to their effect. The killing is exhibited after adsorption to specific receptors located on the external surface of sensitive bacteria, by one of the three primary mechanisms: forming channels in the cytoplasmic membrane, degrading cellular DNA/RNA, or inhibiting protein synthesis. Because of their narrow range of activity, it has been proposed that the primary role of bacteriocins is to mediate intraspecific, or population level, interactions [54]. The genetic determinants of most of the bacteriocins are located on the plasmids, apart from few, which are chromosomally encoded [52]. Bacteriocins of *E. coli* are usually called colicins. A relatively high frequency of colicin-encoding plasmids is found in isolates of pathogenic *E. coli* [55], for example, ~80% of O157:H7 enterohemorrhagic *E. coli* strains studied by Bradley and Howard were colicinogenic [56]. Especially microcins have been associated with pathogenic strains [54]. In a collection of *E. coli* strains isolated from skin and soft-tissue infections, 55% of strains possessed microcin M, and 43% possessed microcin H47 [57]. Further, colicin insensitivity among these strains correlated with a higher prevalence of extraintestinal virulence factors [58]. Typical *E. coli* bacteriocins, their receptors, translocation systems, and mode of action are given in **Table 7**.

Bacteriocin	Receptor	Translocation system	Mode of action
ColA	BtuB	Tol	Ion channel
ColB	FepA	Ton	Ion channel
ColD	FepA	Ton	Stops translation
ColE1	BtuB	Tol	Ion channel
ColE2	BtuB	Tol	DNA-endonuclease
ColE3	BtuB	Tol	rRNA-endonuclease
ColE4	BtuB	Tol	rRNA-endonuclease
ColE5	BtuB	Tol	Stops translation
ColE6	BtuB	Tol	rRNA-endonuclease
ColE7	BtuB	Tol	DNA-endonuclease
ColE8-J	BtuB	Tol	DNA-endonuclease
ColIa	Cir	Ton	Ion channel
ColIb	Cir	Ton	Ion channel
ColK	Tsx	Tol	Ion channel
ColM	FhuA	Ton	Inhibition of peptidoglycan synthesis
ColN	OmpF	Tol	Ion channel
ColS4	OmpW	Tol	Ion channel

Table 7.
Typical E. coli bacteriocins, their receptor, translocation system, and mode of action [59, 60].

3.8 The probiotic *E. coli*

Probiotics are live microorganisms that, when administered in adequate amounts, confer a health benefit on the host. Probiotic bacteria act via a variety of means, including modulation of immune function, production of organic acids and antimicrobial compounds, interaction with resident microbiota, interfacing with the host, improving the gut barrier integrity, and enzyme formation [61]. Several *E. coli* strains were recognized as good and effective probiotics and are now used in drugs (see **Table 8**). The probiotic *E. coli* are applied to a variety of human conditions, including intestinal bowel diseases and diarrhea. Further it was shown that colonization of newborns led to reduced disease rates, lower incidence of allergies, and reduced mortality [62].

E. coli Nissle 1917 is nowadays often used as a reference strain or model microorganism in experimental biomedical studies, including recombinant manipulations of the strain in order to construct derivatives with novel properties [64]. One such example is the strain ŽP, which is a genetically modified Nissle 1917 possessing a bacterial conjugation-based "kill"-"anti-kill" antimicrobial system—a conjugative plasmid carrying the "kill" gene (colicin ColE7 activity gene) and a chromosomally encoded "anti-kill" gene (ColE7 immunity gene). Hence, in the process of conjugation, the conjugative plasmid transfers the "kill" gene into a recipient cell, where it is expressed and the recipient killed [65, 66].

Drug name	Mutaflor	Symbioflor 2	Colinfant newborn
E. coli strain	*E. coli* Nissle 1917 strain	Six different *E. coli* strains (G1/2, G3/10, G4/9, G5, G6/7, and G8)	*E. coli* A0 34/86 strain
Product	Capsules	Suspension	Powder for preparation of per oral solution
Produced by	Ardeypharm GmbH, Herdecke, Germany	SymbioPharm GmbH, Herborn, Germany	Dyntec, Terezín, Czech Republic
Contents	2.5–25 × 10^9 CFU/capsule	1.5–4.5 × 10^7 CFU/ml	0.8–1.6 × 10^8 CFU/dosis
Recommended daily dose	1–2 capsules/day (2.5–50 × 10^9 CFU)	2–4 ml (3.0–18 × 10^7 CFU)	0.8–1.6 × 10^8 CFU three times/week
Isolation date of the used strain(s)	1915	1954	Data not available
Serotype	06:K5:H1	Variable including 035,129, 0:169, rough, all H–	083:K24:H31
Plasmid content	2 cryptic plasmids	12 plasmids	No plasmids
Microcin production	Microcin M, H47	Microcin S	Data not available
Motility	Motile (flagella present)	Nonmotile (flagella absent)	Data not available
Closest relatives	CFT073, ABU83972 (UPEC)	K12, ATCC8739 (commensals)	CFT073, 536 (UPEC)
Year of first publication describing the use in humans	1989	1998	1967

Table 8.
Probiotic E. coli drugs [62, 63].

3.9 The "workhorse" *E. coli*

E. coli is known for its fast growing rate in chemically defined media and extensive molecular tools available for different purposes. All these make it an important model organism, which is also called the "workhorse" of molecular biology. Even though *E. coli* lacks many interesting features appreciated in biotechnology, such as growing at extreme temperatures or pH and the capacity to degrade toxic compounds, pollutants, or difficult to degrade polymers, it is much used in biotechnology also [67]. In **Table 9** contributions of *E. coli* to biology, medicine, and industry are listed.

The following recombinant pharmaceuticals were set up to be in vivo synthesized in *E. coli*: insulin, interleukin-2, human interferon-β, erythropoietin, human growth hormone, human blood clotting factors, pegloticase, taxol, and certolizumab. Further, *E. coli* is also used to produce biofuels and industrial chemicals such as phenol, ethanol, mannitol, and a variety of others [68].

Contribution	Authors	Year
Molecular biology, physiology, and genetics		
Elucidation of the genetic code	Crick FH, Barnett L, Brenner S, and Watts-Tobin RJ	1961
DNA replication	Lehman IR, Bessman MJ, Simms ES, and Kornberg A	1958
Transcription	Stevens A	1960
Life cycle of lytic bacteriophages	Ellis EL and Delbrück M	1939
Gene regulation of the *lac* operon	Jacob F and Monod J	1961
Gene regulation of the *ara* operon	Englesberg E, Irr J, Power J, and Lee N	1965
Discovery of restriction enzymes	Linn S and Arber W	1968
Identification of genes controlling antimicrobial drug tolerance in stationary phase	Hu Y and Coates AR	2005
Role of global regulators and nucleotide metabolism in antibiotic tolerance	Hansen S, Lewis K, and Vulić M	2008
Metabolic control of persister formation	Amato SM, Orman MA, and Brynildsen MP	2013
Swarming motility behavior	Harshey RM and Matsuyama T.	1994
Elucidation of the structure and function of ATP synthase	Capaldi RA, Schulenberg B, Murray J, and Aggeler R	2000
Conjugal DNA transfer	Tatum EL and Lederberg J	1947
Evolution		
Random nature of mutation	Luria SE and Delbrück M	1943
Relationship between genomic evolution and adaptation	Barrick JE, Yu DS, Yoon SH, Oh TK, Schneider D, Lenski RE, and Kim JF	2009
Role of adaption, chance, and history in evolution	Travisano M, Mongold JA, Bennet AF, and Lenski RE	1995
Adaptive mutation	Cairns J, Overbaugh J, and Miller S	1988
Role of historical contingency in evolution	Blount ZD, Borland CZ, and Lenski RE	2008
Origin of novel traits	Blount ZD, Barrick JE, Davidson CJ, and Lenski RE	2012
Long-term fitness trajectories	Wiser MJ, Ribeck N, and Lenski RE	2013
Effect of sexual recombination on adaptation	Cooper TF	2007
Predator-prey interactions (bacteriophage)	Chao L and Levin BR	1977
Genetic engineering and biotechnology		
Molecular cloning and recombinant DNA	Cohen S, Chang A, Boyer H, and Helling R	1973
Generating precise deletions and insertions	Link AJ, Phillips D, and Church GM	1997
Gene replacement	Herring CD, Glasner JD, and Blattner FR	2003

Table 9.
Contributions of E. coli to biology, medicine, and industry [68–70].

4. Conclusion

To conclude, *E. coli* is a truly versatile microorganism possessing many facets—it is a well-known commensal bacterium, but some strains can be also pathogenic, even causing mortality, especially if the pathogenic strain acquired multiple resistance genes. However used as a probiotic it can improve health and in it can be employed as a good working "workhorse" in the laboratory as well as in biotechnological settings. The differentiation between commensal and pathogenic strains is not easy, as among the healthy gut microbiota pathogenic strains are hidden, and also commensal strains can become pathogenic due to horizontal gene transfer of mobile genetic elements possessing virulence genes [71]. Even though *E. coli* has been the object of research now for already more than 100 years, its versatility warrants new possibilities for investigation also in the future.

Acknowledgements

The author is thankful to the Slovenian Research Agency (P1-0198) for its funding.

Conflict of interest

The author has no conflict of interest.

Author details

Marjanca Starčič Erjavec
Biotechnical Faculty, University of Ljubljana, Ljubljana, Slovenia

*Address all correspondence to: marjanca.starcic.erjavec@bf.uni-lj.si

References

[1] PubMed-Search Item: *Escherichia coli*, Publication Dates: From 1885/01/01 to 2018/12/31 [Internet]. Available from: https://www.ncbi.nlm.nih.gov/pubmed [Accessed: February 07, 2019]

[2] Escherich T. The intestinal bacteria of the neonate and breast-fed infant. 1885. Reviews of Infectious Diseases. 1988;**10**:1220-1225

[3] List of Prokaryotic Names with Standing in Nomenclature (LPSN). Genus *Escherichia* [Internet]. Available from: http://www.bacterio.net/escherichia.html [Accessed: February 07, 2019]

[4] Escherich T. Die darmbakterien des neugeborenen und säuglinge. Fortschritte der Medizin. 1885;**3**:515-528;547-554

[5] Madigan MT, Martinko JM, Bender KS, Buckley DH, Stahl DA. Brock Biology of Microorganisms. 14th ed. London: Pearson Education, Inc; 2015

[6] Starčič M, Johnson JR, Stell AL, van der Goot J, Hendriks HG, van Vorstenbosch C, et al. Haemolytic *Escherichia coli* isolated from dogs with diarrhea have characteristics of both uropathogenic and necrotoxigenic strains. Veterinary Microbiology. 2002;**85**:361-377

[7] Kaper JB, Nataro JP, Mobley HL. Pathogenic *Escherichia coli*. Nature Reviews. Microbiology. 2004;**2**:123-140

[8] van Elsas JD, Semenov AV, Costa R, Trevors JT. Survival of *Escherichia coli* in the environment: Fundamental and public health aspects. ISME Journal. 2011;**5**:173-183. DOI: 10.1038/ismej.2010.80

[9] Blattner FR, Plunkett G 3rd, Bloch CA, Perna NT, Burland V, Riley M, et al. The complete genome sequence of *Escherichia coli* K-12. Science. 1997;**277**:1453-1462

[10] GenBank NC_000913.3 [Internet]. 2018. Available from: http://www.ncbi.nlm.nih.gov/nuccore/NC_000913.3 [Accessed: August 03, 2019]

[11] Cheong WH, Tan YC, Yap S, Ng KP. ClicO FS: An interactive web-based service of Circos. Bioinformatics. 2015;**31**:3685-3687. DOI: 10.1093/bioinformatics/btv433

[12] Krzywinski M, Schein J, Birol I, Connors J, Gascoyne R, Horsman D, et al. Circos: An information aesthetic for comparative genomics. Genome Research. 2009;**19**:1639-1645. DOI: 10.1101/gr.092759.109

[13] Schmidt H, Hensel M. Pathogenicity islands in bacterial pathogenesis. Clinical Microbiology Reviews. 2004;**17**(1):14-56

[14] Genome-Search Item *Escherichia coli*. [Internet]. Available from: https://www.ncbi.nlm.nih.gov/genome/167 [Accessed: April 04, 2019]

[15] Frost L, Ippen-Ihler K, Skurray RA. Analysis of the sequence and gene products of the transfer region of the F sex factor. Microbiological Reviews. 1994;**58**:162-210

[16] Firth N, Ippen-Ihler K, Skurray RA. Structure and function of the F factor and mechanism of conjugation. In: Neidhardt FC, editor in chief. Curtiss R III, Ingraham JL, Lin ECC, Low KB, Magasanik B, Reznikoff WS, et al., editors. *Escherichia coli* and *Salmonella*. 2nd ed. Washington: ASM Press; 1996. p. 2377-2401

[17] Saadi S, Maas WK, Bergquist PL. RepFIC, a basic replicon of IncFI plasmids that has homology with a basic

replicon of IncFII plasmids. Plasmid. 1984;**12**:61-64

[18] GenBank AP001918.1 [Internet]. 2018. Available from: https://www.ncbi.nlm.nih.gov/nuccore/ap001918 [Accessed: April 04, 2019]

[19] Chaudhuri RR, Henderson IR. The evolution of the *Escherichia coli* phylogeny. Infection, Genetics and Evolution. 2012;**12**:214-226

[20] Ørskov F, Ørskov I. *Escherichia coli* serotyping and disease in man and animals. Canadian Journal of Microbiology. 1992;**38**:699-704

[21] Milkman R. Electrophoretic variation in *Escherichia coli* from natural sources. Science. 1973;**182**:1024-1026

[22] Herzer PJ, Inouye S, Inouye M, Whittam TS. Phylogenetic distribution of branched RNA-linked multicopy single-stranded DNA among natural isolates of *Escherichia coli*. Journal of Bacteriology. 1990;**172**:6175-6181

[23] Clermont O, Bonacorsi S, Bingen E. Rapid and simple determination of the *Escherichia coli* phylogenetic group. Applied and Environmental Microbiology. 2000;**66**:4555-4558

[24] Walk ST, Alm EW, Gordon DM, Ram JL, Toranzos GA, Tiedje JM, et al. Cryptic lineages of the genus *Escherichia*. Applied and Environmental Microbiology. 2009;**75**:6534-6544. DOI: 10.1128/AEM.01262-09

[25] Luo C, Walk ST, Gordon DM, Feldgarden M, Tiedje JM, Konstantinidis KT. Genome sequencing of environmental *Escherichia coli* expands understanding of the ecology and speciation of the model bacterial species. Proceedings of the National Academy of Sciences of the United States of America. 2011;**108**:7200-7205. DOI: 10.1073/pnas.1015622108

[26] Clermont O, Christenson JK, Denamur E, Gordon DM. The Clermont *Escherichia coli* phylo-typing method revisited: Improvement of specificity and detection of new phylo-groups. Environmental Microbiology Reports. 2013;**5**:58-65. DOI: 10.1111/1758-2229.12019

[27] Logue CM, Wannemuehler Y, Nicholson BA, Doetkott C, Barbieri NL, Nolan LK. Comparative analysis of phylogenetic assignment of human and avian ExPEC and fecal commensal *Escherichia coli* using the (previous and revised) Clermont phylogenetic typing methods and its impact on avian pathogenic *Escherichia coli* (APEC) classification. Frontiers in Microbiology. 2017;**8**:283. DOI: 10.3389/fmicb.2017.00283

[28] Starčič Erjavec M, Predojević L, Žgur-Bertok D. Commentary: Comparative analysis of phylogenetic assignment of human and avian ExPEC and fecal commensal *Escherichia coli* using the (previous and revised) Clermont phylogenetic typing methods and its impact on avian pathogenic *Escherichia coli* (APEC) classification. Frontiers in Microbiology. 2017;**8**:1904. DOI: 10.3389/fmicb.2017.01904

[29] Iranpour D, Hassanpour M, Ansari H, Tajbakhsh S, Khamisipour G, Najafi A. Phylogenetic groups of *Escherichia coli* strains from patients with urinary tract infection in Iran based on the new Clermont phylotyping method. BioMed Research International. 2015;**2015**:846219. DOI: 10.1155/2015/846219

[30] Houghteling PD, Walker WA. Why is initial bacterial colonization of the intestine important to infants' and children's health? Journal of Pediatric Gastroenterology and Nutrition. 2015;**60**:294-307. DOI: 10.1097/MPG.0000000000000597

[31] Stecher B, Hardt WD. Mechanisms controlling pathogen colonization of the gut. Current Opinion in Microbiology. 2011;**14**:82-91. DOI: 10.1016/j.mib.2010.10.003

[32] Apperloo-Renkema HZ, van der Waaij BD, van der Waaij D. Determination of colonization resistance of the digestive tract by biotyping of *Enterobacteriaceae*. Epidemiology and Infection. 1990;**105**:355-361

[33] Robins-Browne RM, Holt KE, Ingle DJ, Hocking DM, Yang J, Tauschek M. Are *Escherichia coli* pathotypes still relevant in the era of whole-genome sequencing? Frontiers in Cellular and Infection Microbiology. 2016;**6**:141. DOI: 10.3389/fcimb.2016.00141

[34] Webb SA, Kahler CM. Bench-to-bedside review: Bacterial virulence and subversion of host defences. Critical Care. 2008;**12**(6):234. DOI: 10.1186/cc7091

[35] Peterson JW. Bacterial pathogenesis. In: Baron S, editor. Medical Microbiology. 4th ed. Galveston: University of Texas Medical Branch at Galevston; 1996. pp. 121-135. Available from: https://www.ncbi.nlm.nih.gov/books/NBK8526/

[36] Salyers AA, Whitt DD. Bacterial Pathogenesis. A Molecular Approach. Washington, D.C.: ASM Press; 1994

[37] Moll A, Manning PA, Timmis KN. Plasmid-determined resistance to serum bactericidal activity: A major outer membrane protein, the *traT* gene product, is responsible for plasmid-specified serum resistance in *Escherichia coli*. Infection and Immunity. 1980;**28**(2):359-367

[38] Cirl C, Wieser A, Yadav M, Duerr S, Schubert S, Fischer H, et al. Subversion of toll-like receptor signaling by a unique family of bacterial toll/interleukin-1 receptor domain-containing proteins. Nature Medicine. 2008;**14**(4):399-406. DOI: 10.1038/nm1734

[39] Aslam B, Wang W, Arshad MI, Khurshid M, Muzammil S, Rasool MH, et al. Antibiotic resistance: A rundown of a global crisis. Infection and Drug Resistance. 2018;**11**:1645-1658. DOI: 10.2147/IDR.S173867

[40] Peterson E, Kaur P. Antibiotic resistance mechanisms in bacteria: Relationships between resistance determinants of antibiotic producers, environmental bacteria, and clinical pathogens. Frontiers in Microbiology. 2018;**9**:2928. DOI: 10.3389/fmicb.2018.02928

[41] ECDC. Surveillance Report: Antimicrobial Resistance Surveillance in Europe 2009. 2010. Available from: https://ecdc.europa.eu/en/publications-data/antimicrobial-resistance-surveillance-europe-2009

[42] ECDC. Surveillance Report: Antimicrobial Resistance Surveillance in Europe 2010. 2011. Available from: https://ecdc.europa.eu/en/publications-data/antimicrobial-resistance-surveillance-europe-2010

[43] ECDC. Surveillance Report: Antimicrobial Resistance Surveillance in Europe 2011. 2012. Available from: https://ecdc.europa.eu/en/publications-data/antimicrobial-resistance-surveillance-europe-2011

[44] ECDC. Surveillance Report: Antimicrobial Resistance Surveillance in Europe 2012. 2013. Available from: https://ecdc.europa.eu/en/publications-data/antimicrobial-resistance-surveillance-europe-2012

[45] ECDC. Surveillance Report: Antimicrobial Resistance Surveillance in Europe 2013. 2014. Available from:

https://ecdc.europa.eu/en/publications-data/antimicrobial-resistance-surveillance-europe-2013

[46] ECDC. Surveillance Report: Antimicrobial Resistance Surveillance in Europe 2014. 2015. Available from: https://ecdc.europa.eu/en/publications-data/antimicrobial-resistance-surveillance-europe-2014

[47] ECDC. Surveillance Report: Antimicrobial Resistance Surveillance in Europe 2015. 2017. Available from: https://ecdc.europa.eu/en/publications-data/antimicrobial-resistance-surveillance-europe-2015

[48] ECDC. Surveillance Report: Antimicrobial Resistance Surveillance in Europe 2016. 2017. Available from: https://ecdc.europa.eu/en/publications-data/antimicrobial-resistance-surveillance-europe-2016

[49] ECDC. Surveillance Report: Surveillance of Antimicrobial Resistance in Europe 2017. 2018. Available from: https://ecdc.europa.eu/en/publications-data/surveillance-antimicrobial-resistance-europe-2017

[50] Tyson GH, McDermott PF, Li C, Chen Y, Tadesse DA, Mukherjee S, et al. WGS accurately predicts antimicrobial resistance in *Escherichia coli*. The Journal of Antimicrobial Chemotherapy. 2015;**70**(10):2763-2769. DOI: 10.1093/jac/dkv186

[51] Cavera VL, Arthur TD, Kashtanov D, Chikindas ML. Bacteriocins and their position in the next wave of conventional antibiotics. International Journal of Antimicrobial Agents. 2015;**46**(5):494-501. DOI: 10.1016/j.ijantimicag.2015.07.011

[52] Daw MA, Falkiner FR. Bacteriocins: Nature, function and structure. Micron: The International Research and Review Journal for Microscopy. 1996;**27**(6):467-479

[53] Duquesne S, Destoumieux-Garzón D, Peduzzi J, Rebuffat S. Microcins, gene-encoded antibacterial peptides from enterobacteria. Natural Product Reports. 2007;**24**(4):708-734

[54] Riley MA, Gordon DM. The ecological role of bacteriocins in the bacterial competition. Trends in Microbiology. 1999;**7**:129-133

[55] van der Wal FJ, Luirink J, Oudega B. Bacteriocin release proteins: Mode of action, structure, and biotechnological application. FEMS Microbiology Reviews. 1995;**17**:318-399

[56] Bradley DE, Howard SP. Colicinogeny of O157:H7 enterohemorrhagic *Escherichia coli* and the shielding of colicin and phage receptors by their O-antigenic side chains. Canadian Journal of Microbiology. 1991;**37**:97-104

[57] Starčič Erjavec M, Petkovšek Ž, Predojević L, Žgur-Bertok D. Iron uptake and bacteriocin genes among *Escherichia coli* strains from skin and soft-tissue infections. APMIS (Acta Pathologica, Microbiologica et Immunologica Scandinavica). 2017;**125**(3):264-267. DOI: 10.1111/apm.12663

[58] Petkovšek Ž, Žgur-Bertok D, Starčič Erjavec M. Colicin insensitivity correlates with a higher prevalence of extraintestinal virulence factors among *Escherichia coli* isolates from skin and soft-tissue infections. Journal of Medical Microbiology. 2012;**61**(Pt 6):762-765. DOI: 10.1099/jmm.0.037234-0

[59] Arnold T, Zeth K, Linke D. Structure and function of colicin S4, a colicin with a duplicated receptor-binding domain. The Journal of Biological Chemistry. 2009;**284**(10):6403-6413. DOI: 10.1074/jbc.M808504200

[60] Cascales E, Buchanan SK, Duché D, Kleanthous C, Lloubès R, Postle K, et al. Colicin biology. Microbiology and

Molecular Miology Reviews: MMBR. 2007;**71**(1):158-229

[61] Sanders ME, Merenstein DJ, Reid G, Gibson GR, Rastall RA. Probiotics and prebiotics in intestinal health and disease: From biology to the clinic. Nature Reviews. Gastroenterology & Hepatology. 2019. DOI: 10.1038/s41575-019-0173-3

[62] Wassenaar TM. Insights from 100 years of research with probiotic *E. coli*. European Journal of Microbiology and Immunology. 2016;**6**(3):147-161. DOI: 10.1556/1886.2016.00029

[63] Wassenaar TM, Zschüttig A, Beimfohr C, Geske T, Auerbach C, Cook H, et al. Virulence genes in a probiotic *E. coli* product with a recorded long history of safe use. European Journal of Microbiology and Immunology. 2015;**5**:81-93. DOI: 10.1556/EuJMI-D-14-00039

[64] Sonnenborn U. *Escherichia coli* strain Nissle 1917-from bench to bedside and back: History of a special *Escherichia coli* strain with probiotic properties. FEMS Microbiology Letters. 2016;**363**(19):pii: fnw212. DOI: 10.1093/femsle/fnw212

[65] Starčič Erjavec M, Petkovšek Ž, Kuznetsova MV, Maslennikova IL, Žgur-Bertok D. Strain ŽP: The first bacterial conjugation-based "kill"- "anti-kill" antimicrobial system. Plasmid. 2015;**82**:28-34. DOI: 10.1016/j.plasmid.2015.10.001

[66] Maslennikova IL, Kuznetsova MV, Toplak N, Nekrasova IV, Žgur Bertok D, Starčič Erjavec M. Estimation of the bacteriocin ColE7 conjugation-based "kill"- "anti-kill" antimicrobial system by real-time PCR, fluorescence staining and bioluminescence assays. Letters in Applied Microbiology. 2018;**67**(1):47-53. DOI: 10.1111/lam.12884

[67] Vargas-Maya NI, Franco B. *Escherichia coli* as a model organism and its application in biotechnology. In: Samie A, editor. *Escherichia coli*: Recent Advances on Physiology, Pathogenesis and Biotechnological Applications. Rijeka: InTechOpen; 2017. pp. 253-274. DOI: 10.5772/67306

[68] Blount ZD. The unexhausted potential of *E. coli*. eLife. 2015;**4**:e05826. DOI: 10.7554/eLife.05826

[69] Foster PL. Adaptive mutation: Implications for evolution. BioEssays: News and Reviews in Molecular, Cellular and Developmental Biology. 2000;**22**(12):1067-1074

[70] Tatum EL, Lederberg J. Gene recombination in the bacterium *Escherichia coli*. Journal of Bacteriology. 1947;**53**(6):673-684

[71] Starčič Erjavec M, Žgur-Bertok D. Virulence potential for extraintestinal infections among commensal *Escherichia coli* isolated from healthy humans: The Trojan horse within our gut. FEMS Microbiology Letters. 2015;**362**(5):pii. fnu061. DOI: 10.1093/femsle/fnu061

Section 2

Versatility of *Escherichia coli*

Chapter 2

Escherichia coli: A Versatile Platform for Recombinant Protein Expression

Wan-Keung Raymond Wong, Ka-Lun Ng, Thiyagarajan Sivakumar, Xiu-Hua Hu, Hao Wang and Lai-Cheuk Nelson Lai

Abstract

Among the living organisms, *Escherichia coli* has been the most common choice employed for recombinant protein expression. In addition to its well-characterized genetics, *E. coli* is fast growing, relatively cheap, and easy to handle. These fine properties, in conjunction with the success achieved in transforming plasmid DNA into *E. coli*, as well as the advent of various genetic engineering techniques in the 1970s, have enabled *E. coli* to be considered as the most favorable host for genetic manipulations. The recent advances in better comprehension of regulatory controls of gene expression and the availability of various novel approaches, which include both intracellular, e.g., through intein-mediated expression and self-cleavages, and extracellular, e.g., through the use of secretion signals, to achieve successful expression of the target proteins in *E. coli* further support the view that *E. coli* is the most promising host choice for heterologous protein expression.

Keywords: *Escherichia coli*, *E. coli*, recombinant protein expression, heterologous protein, authentic structures, fusion protein, affinity tags, secretion, excretion, inteins

1. Introduction

The achievements in unveiling the structure of DNA, deciphering the genetic code, understanding gene expression and regulation, and discovering extrachromosomal DNA (plasmid), restriction endonucleases and DNA ligases in the 1950s and 1960s laid the groundwork for the construction of the first chimeric (recombinant) DNA molecule [1]. In 1973, Cohen and Boyer reported their success in creating the first *Escherichia coli* transformant into which a recombinant plasmid molecule was introduced [2].

The possibility of inserting foreign DNA into *E. coli* has not only allowed the development of a vast number of molecular biology techniques for genetic manipulations, e.g., construction and characterization of cDNA libraries, DNA splicing and amplification, hybridization and sequencing, site-specific mutagenesis, research and applications of bacteriophages and DNA modifying enzymes, studies of regulation of gene expression, etc., but also the exploitation of *E. coli* for use as a surrogate host for the expression of heterologous proteins. Despite the presence of restriction systems in *E. coli* [3, 4], foreign DNA from a wide variety of sources, ranging from a simple virus to a complex organism such as the human being, have been shown to be stably maintained in it. Therefore, although *E. coli* may not be employed as the final host cell for target protein expression, due to the difficulties of DNA manipulations, which include selection, characterization, and amplification of recombinant DNA constructs directly performed in the final host cell, *E. coli* is routinely recruited to fulfill the tasks. In this regard, *E. coli* acts as an indispensable stepping stone facilitating molecular genetic studies undertaken in eukaryotic and other bacterial host cells.

The fine properties including fast growth rate (with a doubling time of less than 30 min), ease of manipulation, relatively low cost of cultivation/protein expression, and the well-characterized genetics of *E. coli* are the key reasons why this bacterium has been commonly involved in the wide range of studies mentioned above. Thus, *E. coli* has been crowned the most popular host of choice employed for heterologous protein expression.

Being a Gram-negative bacterium, *E. coli* possesses two layers of cell envelope, the inner/cytoplasmic membrane and the outer membrane, between which is the narrow periplasmic space (**Figure 1**). Except for specific strains, e.g., a colicin-producing strain, *E. coli* does not produce proteins beyond the outer membrane in the living environment. Therefore, in the early days, *E. coli* was essentially employed for cytoplasmic/intracellular expression of recombinant proteins. However, the advancement of recombinant DNA technology in the 1970s–1980s not only enabled *E. coli* to serve as an efficient host system for protein expression but also facilitated *E. coli* to be transformed into a highly versatile expression platform, with which heterologous proteins may also be allowed to be produced in the culture supernatant. In addition, despite being expressed in the cytoplasm, target proteins resulting from a recently developed novel process, the intein-mediated expression approach [5, 6], have been shown to possess primary structures precisely the same as expected [7]. This new recombinant DNA approach may prove to be a practical

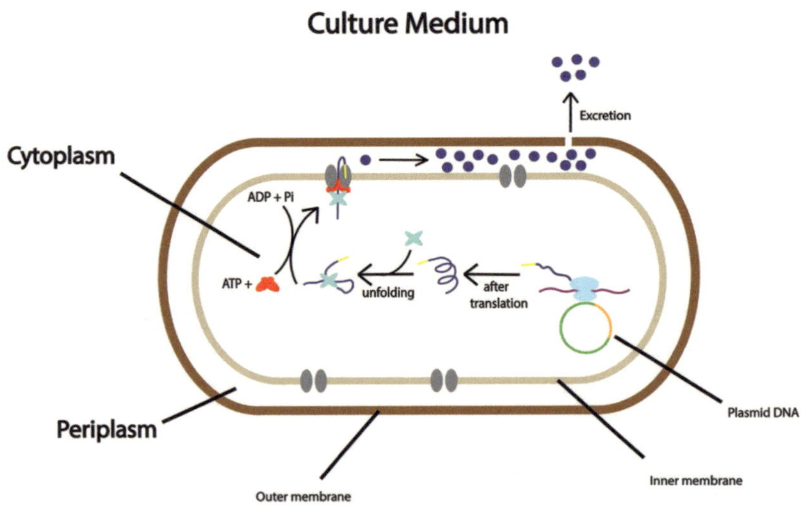

Figure 1.
Schematic representation of the subcellular compartments of E. coli available for expression of heterologous proteins. These compartments such as cytoplasm, periplasm, and culture medium are shown. For a secretory protein, it is initially formed as a preprotein ( ), which consists of a signal peptide ( ). The preprotein is directed to the SecYEG pathway. Before reaching the secretion channel ( ), the preprotein is unfolded by the SecB protein ( ), followed by interacting with the SecA protein ( ), which helps hydrolyze ATP to support the active transport of the preprotein. The signal peptide is then cleaved off from the preprotein subsequent to the passage of the latter through the SecYEG pathway.

process for the recombinant production of medically valuable proteins that are preferred to share the authentic structures with their native counterparts.

Proteins expressed in eukaryotic cells are subjected to post-translational modifications (PTM), of which many of them do not appear to occur in *E. coli* cells. Despite this deficiency, *E. coli* is still the most common choice employed for the expression of eukaryotic proteins. For example, about 30% of the medically valuable proteins produced using recombinant DNA approaches are expressed employing *E. coli* as the host [8]. The finding of successful expression of eukaryotic proteins in E. coli suggests that many target proteins may not be posttranslationally modified, and even some of them are, PTM may not have a direct effect on functional activities. The observation also supports the view that *E. coli* will continuously play an indispensable role in heterologous protein expression. In choosing the most appropriate tactic for the expression of a heterologous protein in *E. coli*, it is important that we understand both the target protein and the available methods of choice well. *E. coli* is recognized as being a "versatile" host from the perspective that it may facilitate heterologous protein expression in all three of the subcellular compartments including: (i) cytoplasm, (ii) periplasm, and (iii) culture medium (**Figure 1**). In this communication, we discuss how those compartments may be employed to express foreign proteins that share widely different biochemical properties, under the condition that the presence of PTM is not a prerequisite.

2. Expression of recombinant proteins in the cytoplasm of *E. coli*

The breakthrough achievements in the construction of recombinant DNA molecules [1] and the transformation of chimeric DNA constructs into *E. coli* [2] in the early 1970s have paved the way for rapid advances in the development of recombinant DNA approaches to the expression of a wide collection of useful/valuable proteins of various origins. Due to the aforementioned fine properties of *E. coli*, this Gram-negative bacterium has been extensively studied and exploited for use as a host to facilitate expression of heterologous proteins (**Table 1**).

2.1 Fusion protein approach

A common strategy in the expression of heterologous proteins is to fuse the target protein with a fusion partner, of which a familiar example is the enzyme β-galactosidase (β-Gal) expressed intracellularly in *E. coli*. Being well-characterized in terms of its structure and regulation of expression [27, 28], in the early days, β-Gal was one of the few *E. coli* products to be employed as a reporter protein, for which convenient detection assays [29] were available. Fusing the short mammalian somatostatin (Som) comprising only 14 amino acids (aa) to β-Gal, in 1977, Itakura et al. demonstrated (**Figure 2**), for the first time, successful expression of bioactive recombinant somatostatin in *E. coli* [9]. In the work, Som was fused to β-Gal through the application of oligonucleotide assembly. Thus, expression of the two proteins, which was under the regulatory controls of the Lac operon, would result initially in a β-Gal-Som precursor. The β-Gal component played two important roles in the fusion: first, it offered a facile screening assay for the selection of potentially positive clones expressing β-Gal-Som; second, it served as a guardian protecting Som from being attacked by proteolytic degradation from the N-terminus.

Since Som is a short polypeptide consisting of only 14 aa residues [30], which does not consist of Met as a member, in engineering the aforementioned β-Gal-Som fusion, a Met residue was intentionally inserted precisely between β-Gal and Som, thus resulting in a β-Gal-Met-Som precursor in the work [9]. In vitro cleavage with

Approach	Examples	Section concerned	References
Fusion expression	Hormone somatostatin	2.1	[9, 10]
	Human insulin		
Direct expression	Human growth hormone	2.2	[11–15]
	Human fibroblast interferon		
	Human leukocyte interferon (LeIF A)		
	Human leukocyte interferon (LeIF B)		
	Mouse granulocyte-macrophage colony stimulating factor (mGM-CSF)		
Tagged protein expression	Human insulin-like growth factor I (IGF-1)	2.3	[16–18]
	Human parathyroid hormone (hPTH)		
	Human fibroblast growth factor 21 (hFGF21)		
Intein-mediated	Cre recombinase	3	[5, 6, 19, 20]
	Cecropin		
	Human basic fibroblast growth factor (bFGF)		
Secretion/excretion	Human epidermal growth factor (Secretion)	4	[21–26]
	Hirudin		
	Human growth hormone (hGH)		
	Human granulocyte-macrophage colony stimulating factor (Hgm-CSF)		
	Human epidermal growth factor (Excretion)		

Table 1.
Various approaches for heterologous proteins expression in E. coli.

the chemical, CNBr, which specifically attacks Met, resulted in the separation of the fusion to yield authentic Som comprising 14 aa as the final product, which was subsequently shown to be bioactive (**Figure 2**) [9].

A similar approach was also applied to express bioactive human insulin in *E. coli* in 1979 (**Figure 2**), once again, taking advantage of β-Gal as the fusion partner and the absence of a Met residue in the polypeptide [10].

However, the above two examples appear to be the exception rather than the rule. Despite the application of β-Gal to serving as a fusion partner in many other cases of recombinant protein expression, due to the presence of one or more Met residues in the target proteins, the intriguing tactic of employing CNBr to cleave the fusion precursors to free the desired final protein to be impractical for routine use.

2.2 Direct expression

If a heterologous protein is insusceptible to proteolytic degradation by *E. coli* proteases, perhaps a simple method to achieve expression of the protein in *E. coli* is to clone its gene determinant downstream from a regulatory region comprising both the promoter and RBS sequences carried on a suitable expression vector. Examples

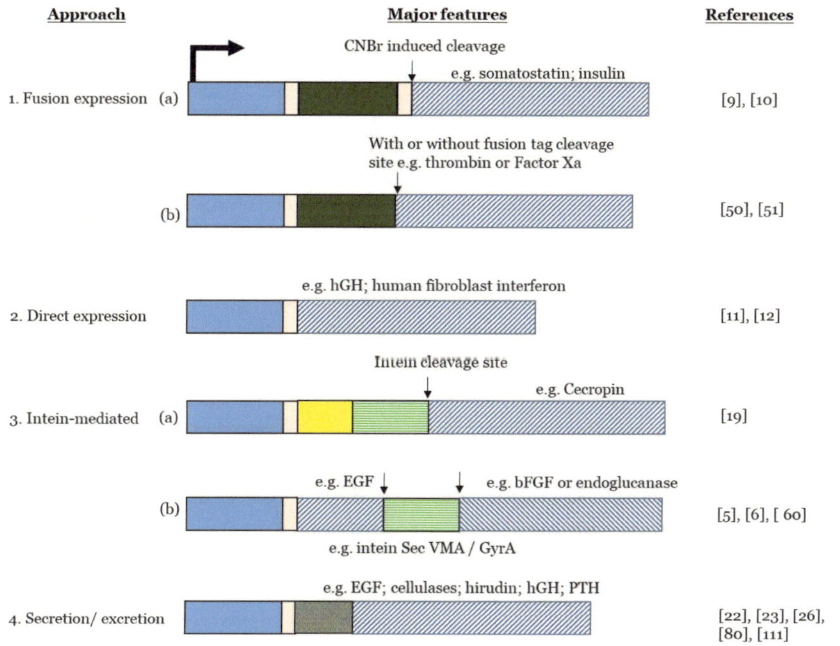

Figure 2.
Schematic diagram depicting the major components included in various expression constructs for recombinant protein expression in E. coli. Regulation region (▭) includes the promoter (↱), operator, and ribosome binding site. Other components including the ATG start codon (▯), and coding sequences for fusion tag (▬), e.g., β-Gal in Approach 1(a), GST, 6xHis-tag, or MBP in Approach 1(b); signal peptide (▭) in Approach 4; chitin-binding domain (▭); target proteins (▨/▧); intein (▤) are also shown. Cleavage sites on fusion proteins (Approach 1a and 1b) and intein-target protein precursors are indicated (↓).

including human growth hormone [11], human hemoglobin [31], interleukin [32], etc., have been expressed in *E. coli* using this approach (**Figure 2**).

The alignment between the target gene and its expression regulatory elements could be conveniently achieved using site-directed mutagenesis. However, the translation initiator, N-formyl-methionine (fMet), which is present in proteins formed in bacteria, may cause adverse effects on the bioactivity and stability of the target protein [33]. The efficiency of removal of fMet in the cell is incomplete and is highly dependent on the adjacent two residues next to the initiator [34, 35].

Various strategies have been described to remove fMet from heterologous proteins including the use of both in vivo and in vitro approaches [34, 36, 37]. However, none of the available protocols is able to result in a homogeneous product that is free of fMet [34, 36]. The target protein, being contaminated by the presence of the undesirable fMet-bearing variant, may exhibit increased immunogenicity [33] and reduced levels of stability and bioactivity [34], which might have a correlation with fMet which has been speculated to serve as a degradation signal [38].

2.3 Applications of affinity tags

A major goal in recombinant DNA expression is to achieve efficient production of a target protein on a large scale. Common strategies including the use of: (1) plasmids with increased copy numbers such as the ones employing runaway replicons [39, 40]; (2) strong transcriptional control signals including P_L, Tac, and T7 promoters [41, 42]; (3) efficient ribosome binding sites such as the Shine-Dalgarno sequence [43]; (4) inducible promoters which may be activated by heat shocking [44], light induction [45] or chemicals, e.g., isopropyl β-D-1-thiogalactopyranoside (IPTG); (5) a codon-optimized gene sequence [46, 47]; and (6) an efficient plasmid maintenance system. These various methods have been commonly applied, either individually or in conjunction with a fusion approach, to achieving efficient expression of target proteins in *E. coli*.

Although high yields of products may result from the application of above mentioned expression approaches, oftentimes, the products present themselves as insoluble inclusion bodies or aggregates. Unfortunately, these inclusion bodies are composed of denatured and misfolded proteins, which are functionally inactive [48, 49]. Due to the rearrangements of disulfide bridges in the aggregates, despite going through the processes of denaturation and renaturation, the target proteins are unlikely able to regain their functional activities [48].

Fusion of a target protein to an affinity tag presents a viable approach to not only the purification of the final product, but also the preservation of the product as a soluble protein. It has been shown that protein tags such as maltose-binding protein [50], glutathione S-transferase [51], small ubiquitin modifying protein [52, 53], and thioredoxin [54] might help improve the solubility of fusion products formed between the tags and target proteins (**Figure 2**). Given that a fusion product is expressed as a soluble and properly folded intermediate, and that it is readily purified using affinity chromatography and proteolytically processed at a recognition site engineered between the tag and target proteins, the frailty of this fusion approach is how the affinity tag may be removed from the target protein on condition that the latter till possesses the peptide sequence as stipulated. Thus, this approach may not be able to meet the stringent demand from therapeutic proteins of which any discrepancy found in their primary structures may result in undesirable side effects such as increased immunogenicity [33], reduced levels of stability and bioactivity [34, 55, 56], and worse still, greater tendency to promote malignancy. It is believed that target proteins bearing the authentic structures are as safe as their native counterparts in performing biological functions [57].

3. Intein-mediated expression of heterologous proteins

3.1 Inteins as fusion partners

Since the first intein, or protein intron, was discovered in the late 1980s [58], over 600 putative intein genes have been discovered [59]. Being able to undergo autocatalytic cleavages of themselves from sequences flanking their two termini, the N- and C-exteins, the application of inteins to the development of *E. coli* expression platforms has revolutionized the production of recombinant proteins in two different facets. First, fusion proteins formed between inteins and target proteins may undergo auto-cleavage activities in the cytoplasm of *E. coli* [5, 6, 60–63]. Second, despite taking place intracellularly, the detached target proteins possess the requisite structures, e.g., the authentic N-terminal sequences which are the same as those of their native counterparts [5, 6, 60–63].

3.2 Autocatalytic cleavages of intein-target fusion proteins: through an in vitro method

In the early days of exploiting the application of inteins to protein expression, fusion precursors formed among three components, comprising an N-terminal protein tag, a common example being a chitin-binding domain (CBD; [64]), a central intein, and a C-terminal target protein (CBD-I-TP), were frequently expressed as biologically inactive inclusion bodies in the cytoplasm of *E. coli* [65–67]. Subsequent to denaturation and renaturation of the protein aggregates [67], the renatured precursors comprising CBD and the target proteins, e.g., Cre recombinase, α-1-antitrypsin, human epidermal growth factor [67] (**Figure 2**), collected in a chitin column was cleaved [64] by modulating the environmental conditions to release the target proteins [67] (**Figure 3**).

Being expressed as inclusion bodies, as discussed in Section 2.3, it is unlikely that the renatured CBD-I-TP molecules would all be bound to the chitin matrix or be correctly refolded. Therefore, the above described intein-mediated expression process working in conjunction with an in vitro autocatalytic cleavage protocol is expected to result in a substantial loss of bioactive target proteins.

3.3 Autocatalytic cleavages of intein-target fusion proteins: in vivo

Despite the inducibility of self-cleavages of inteins by modulating the environmental conditions [68], the exact mechanisms regarding how the induction works is not clear. Recent findings have shown that the ability of an intein element in fusion proteins to undergo self-cleavages appears to be dependent upon the presence of a pair of "well-matched" heterologous "exteins." If this condition is fulfilled, autocatalytic cleavages might take place at the two terminal junctions where the intein is fused with the two exteins (**Figure 3**). It was demonstrated that when human epidermal growth factor (EGF) and basic fibroblast growth factor (bFGF) were precisely fused at the N- and C-termini, respectively, of the *Sce* VMA intein, auto-cleavage processing occurred [5] (**Figure 2**). Both EGF and bFGF were retrieved and shown to share not only authentic structures, but also potent bioactivities with their native counterparts [5].

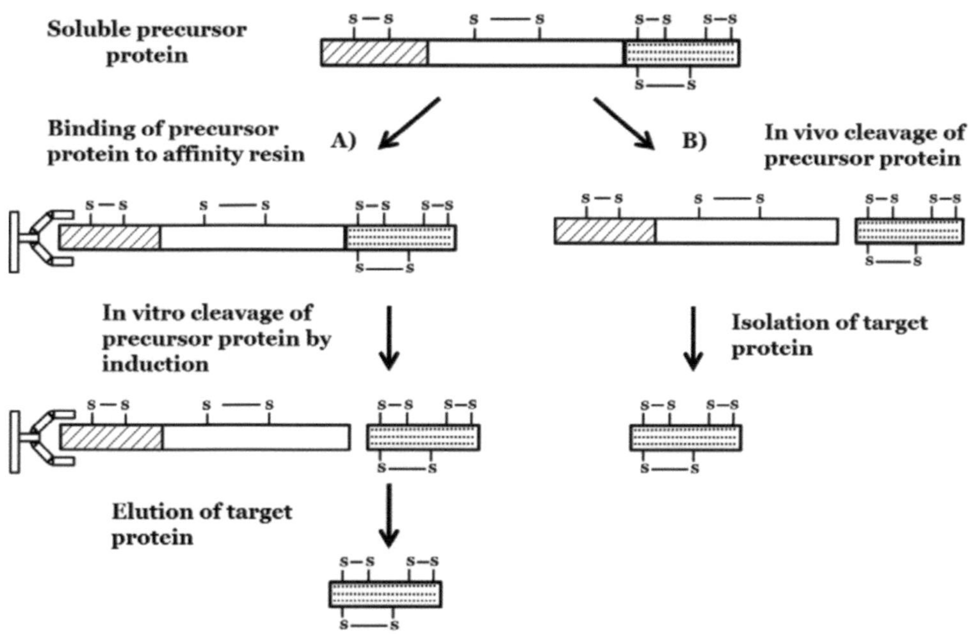

Figure 3.
Processing of precursors formed between inteins and target proteins. The processing may be achieved by one of the following two methods: (A) induced in vitro cleavage; (B) in vivo cleavage. To attain a soluble precursor protein for processing in (A), the insoluble precursor protein express in the host cells is first required to be denatured and renatured. The solubilized precursor is then adsorbed onto an affinity column containing, e.g., chitin resin; changes of the conditions in the column may result in self-cleavage and hence the detachment of the target protein. In (B), the self-cleavage reaction takes place automatically in the host without extra input of effort. In the figure, hatched box ( ) denotes affinity tag, white box ( ) denotes intein, dotted box ( ) denotes target protein and affinity resin is denoted by . The correctly folded affinity tag, intein, and target protein are denoted as , , and, respectively. However, the in vitro cleavage process shown in (A) may also result in incorrectly folded proteins.

Moreover, since EGF was fused to the OmpA signal peptide (OmpA) in the abovementioned work (**Figure 2**), the EGF-VMA-bFGF fusion was also shown to be secretory and both EGF and bFGF were finally detected to be present in the culture medium of their *E. coli* host [5]. Interestingly, when EGF was absent in the fusion, thus leaving the formation of OmpA-VMA-bFGF, and when the positions of EGF and bFGF in the fusion were switched, thus leading to the expression of OmpA-bFGF-VMA-EGF, neither of the two precursors resulted in successful self-cleavages to yield authentic bFGF as the final product [5]. The results support the idea that not only the presence of a matched pair of exteins, but also their relative position in the fusion is important in effecting autocatalytic cleavages of the extein from their intein fusion partner.

Another noteworthy observation from the above work is the soluble nature of the fusion precursor, EGF-VMA-bFGF. This unusual condition, which contrasts markedly with the results of insoluble aggregates reported previously [64], has facilitated auto-cleavages of the fusion precursor to undergo self-cleavages directly in the cytoplasm, thereby avoiding the involvement of a time-consuming and ineffective process of denaturation and renaturation, followed by the extra time and effort spent on implementing the in vitro cleavage operation (**Figure 3**).

The in vivo autocatalytic processing approach introduced above may also be extended for use in the co-expression of other target proteins [60] (**Figure 4**). Moreover, through a combined protocol of gene amplification and refined fed-batch fermentation, the EGF-VMA-bFGF fusion has been upgraded to result in an expression of EGF-VMA-bFGF-VMA-bFGF as the precursor in *E. coli* [6]. Despite 92% bigger in size than EGF-VMA-bFGF, which was shown to have a mass of 73 kDa [5], EGF-VMA-bFGF-VMA-bFGF was found to be expressed as a soluble protein, which was still able to undergo autocatalytic cleavages to result in authentic and bioactive

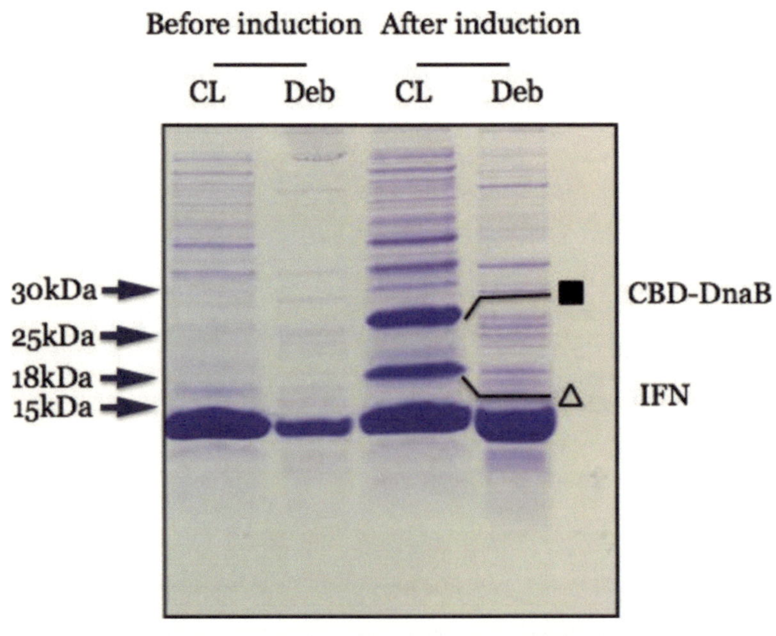

Figure 4.
Analysis of the CBD-DnaB-IFN precursor by polyacrylamide gel electrophoresis. Cell lysate samples were prepared from E. coli transformants harboring plasmid pTWIN-CBD-DnaB-IFN grown in 2×YT medium before and after 4 h of IPTG induction. Cell pellets were lysed by a chemical lysis protocol as described previously [5]. Lanes containing soluble proteins retrieved from the clear lysate (CL) and insoluble proteins retrieved from cell debris (Deb) were resolved as shown. The products CBD-DnaB (■) and IFN (Δ) resulting from autocatalytic cleavages of CBD-DnaB-IFN are denoted.

bFGF as the final product [6, 61–63]. In addition, fermentative production of EGF-VMA-bFGF-VMA-bFGF resulted in a dramatic improvement in the yield bFGF, amounting to 610 mg L^{-1} of cell culture [6], which was over 2.4 times higher than that resulting from the processing of EGF-VMA-bFGF expressed previously [5].

4. Expression of heterologous proteins across the inner membrane of *E. coli*

The approach of secretory expression of heterologous proteins stemmed from the work of W. Gilbert's group, which employed the N-terminal 23 amino acid leader sequence of the *E. coli* penicillinase [69], to direct secretion of eukaryotic proteins, using rat proinsulin as the model protein, to the periplasmic space of *E. coli* in the late 1970s and early 1980s [69–72]. The secreted proinsulin was not only shown to possess an authentic structure, with the cleavage of the signal peptide done precisely [71], but also shown to be more stable than its cytoplasmic counterparts fused to defective signal sequences [72]. Over the next few years, different eukaryotic proteins, e.g., EGF [21], human interferon-α [73], hirudin [22], human growth hormone [23], and human granulocyte-macrophage colony stimulating factor [24], were also successfully expressed though secretion using various bacterial signal peptides in *E. coli* (**Figure 2**).

Meanwhile, *E. coli* mutants that were able to leak endogenous enzymes from the periplasm were isolated [74, 75]. The results suggested that heterologous proteins might also leak from the periplasm to the culture medium in *E. coli*. As expected, a few years later, heterologous proteins including bacterial endoglucanases [76, 77], a penicillinase of an alkalphilic *Bacillus* [78], as well as human proteins, such as β-endorphin [79], EGF, parathyroid hormone, and interleukin-6 [80], were expressed as extracellular products using either wildtype or leaky *E. coli* strains as hosts.

Not all proteins, e.g., the cytoplasmic enzyme—β-galactosidase, may be possibly expressed as secreted or excreted products in *E. coli*, despite their fusions to secretory proteins [81, 82] or directly to the signal peptides of these proteins [83]. Intracellular proteins do not appear to possess a molecular structure that is compatible with the SecYEG pathway, the major translocation machinery located in the inner membrane for protein transport [70, 82, 84–86]. On the other hand, when a naturally secreted target protein, e.g., EGF, is fused to a signal peptide, it may end up as a mature protein in either the periplasm [21] or the culture medium [80] of *E. coli* cells, depending essentially upon the efficiencies of expression and secretion of the protein (see below).

4.1 Secretory expression of target proteins in *E. coli*

In *E. coli*, several protein export systems, including the SecYFG (a trimeric complex comprising three polypeptides: SecY, SecE and SecG), Tat (twin-arginine translocation), and SRP (signal recognition particle) pathways which are embedded in the inner or cytoplasmic membrane, are responsible for the transport of proteins from the cytoplasm to the periplasm [86–89] (**Figure 1**). Among them, the SecYEG translocon is a general, conserved, and essential pathway which is found in both prokaryotic and eukaryotic cells [85, 86]. Being the major protein transport system, over 90% of the translocated proteins are secreted through the SecYEG pathway [87, 88] (**Figure 1**).

To enable proteins to be secreted using the SecYEG translocon, they are required to be expressed first as preproteins, which are fused at their N-termini with a short (commonly less than 24 amino acids) signal peptide [87]. In the cytoplasm, a preprotein is maintained in an extended (export-competent) state by interacting with the SecB chaperone. Subsequent to an interaction formed between the signal and the SecA ATPase, the preprotein-complex then associates with the SecYEG pathway. With repeated pushes of SecA, the preprotein is secreted through the translocon in an ATP-dependent manner, followed by removal of the signal peptide by signal peptidase before the mature protein is released to the periplasm [87, 89, 90].

A wide range of heterologous proteins including degradative enzymes [91, 92], human hormones, and growth factors [5, 25, 26] have been successfully expressed as secretory proteins in *E. coli*. Many of the secreted proteins were not only shown to be bioactive, but also confirmed by sequence determination to possess the correct structures, supporting that the signal peptides fused at the N-termini of the preproteins had been removed correctly during the process of secretion [5, 25]. These advantages, together with lower levels of protease complexity and activity [72], and the relatively more oxidative environment that may help proper folding and disulfide-linkage formation [93, 94], enable the periplasmic space to be consider as a reasonable and appropriate destination for the expression of recombinant secreted proteins. Later on, with the help of various genetic and/or biochemical manipulations [95], or even merely through improving the levels of the secreted proteins concerned, which was referred as the "self-driven approach" [95], interestingly, the target proteins might then to be allowed to leak out to the culture medium, a process termed excretion, which is essentially caused by non-specific leakage of periplasmic proteins (see below).

4.2 Excretory production of target proteins in *E. coli*

In the mid-1980s, researchers from different groups discovered that heterologous proteins expressed and secreted to the periplasm of *E. coli* might also be further excreted to the culture supernatant [22, 79]. For example, the development of sensitive screening assays, e.g., the Congo red plate assay (**Figure 5**), helped to confirm that the detection of a recombinant endoglucanase (Eng) encoded by the *cenA* gene of a Gram-positive bacterium, *Cellulomonas fimi* [77, 91, 96], in the culture supernatant of its *E. coli* host was due to a new phenomenon, excretion (extracellular production), rather than from cell lysis.

Efficient expression regulatory elements such as the strong promoters including *tac*, pL, and T7 [97, 98], the consensus ribosome binding site [99], the coding sequence for the potent OmpA signal [100], an effective inducible system, e.g., the *lac* operator/repressor system for transcriptional regulation [101, 102], etc., which are carried on a stable and high-copy number vector, e.g., pUC18 [103], have been made available to improve not only secretory, but also excretory expression of a wide variety of proteins in *E. coli*. The achievement of this research milestone was well exemplified by the development of an efficient protocol for extracellular production of EGF [104]. In early attempts to express EGF as a secretory protein in *E. coli*, the relatively weak *phoA* promoter was employed to perform transcription of the *egf* gene and the less efficient PhoA signal peptide to direct EGF for secretion. Despite the demonstration of EGF secretion, the EGF detected in the periplasm was only at a considerably low level of 2.4 mg L^{-1} [21]. However, when the *tac* promoter and the *ompA* leader sequence were employed to facilitate EGF expression and secretion, respectively, it was shown that the level of excreted, but not secreted, EGF was markedly improved in *E. coli* cells [80] (**Figure 6**). Moreover, further improvements in EGF expression resulted in a dramatic increase in the yield of excreted EGF [26, 95, 104, 105]. Similar trends were observed in increasing the levels of excretory production of other heterologous proteins, e.g., *C. fimi* cellulases Eng [91, 106, 107] and exoglucanase, as well as bFGF [6].

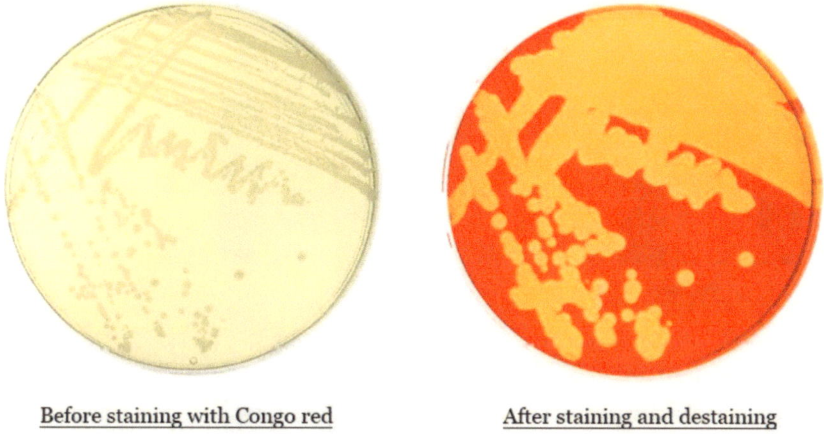

Figure 5.
E. coli transformants expressing excreted Eng. Making use of a convenient and sensitive agar plate assay based on the ability of Eng to degrade carboxymethylcellulose (CMC), and the affinity of the dye Congo red for long molecules of CMC, E. coli cells excreting Eng were detected. In this assay, CMC molecules are degraded randomly by Eng to generate oligomers of the substrate. These oligomers, especially those with fewer than five sugar units, bind poorly to Congo red. Hence, after staining with Congo red and destaining with salt solution, clear zones where CMC has been hydrolyzed by Eng are revealed [110].

4.3 Difficulties in implementing an effective excretory process and potential solutions

The findings described above support the view that excretion is a promising approach for recombinant production of heterologous proteins in *E. coli*. However, it has been shown that not all naturally secreted proteins may be expressed using excretion, despite using efficient transcriptional and translational controls, as well as effective secretion signal. For example, using the same regulatory elements which enabled a high level (325 mg L^{-1}) of excretion of the 53-amino acid (aa) EGF peptide in *E. coli* [26, 95, 104, 105], attempts to produce authentic bFGF (146 residues) by excretion in *E. coli* were unsuccessful [5]. One might wonder whether the discrepancy between the results of EGF and bFGF excretion was due to the marked difference between their molecular sizes. However, cellulases such as Eng and Exg, which possess large mature forms comprising 418 aa [91] and 443 aa [108], respectively, have been shown to be efficiently produced by excretion in *E. coli* [92, 107, 109]. Therefore, in addition to the molecular size of a heterologous protein which might have some effect on the efficiency of excretory production of the protein (see below), it appears that other factor(s), which may be associated to either the protein itself or the host (or both), play a crucial role in determining whether a protein may be expressed as a secretory/excretory product or not.

A major hurdle for excretory production of heterologous products is the dramatic cell death during enhanced expression of the preproteins—the fusion precursors formed between signal peptides and target proteins. A model designated "Saturated Translocation" was proposed to explain the phenomenon of cell lethality resulting from hyper-expression of the preproteins [110]. According to the model, when a preprotein exceeded a tolerable level, it would saturate the capacity of the SecYEG pathway and interfere with its normal function in exporting endogenous proteins. These functional disorders resulted finally in cell death [110]. The model also explained why heterologous proteins of different sizes, which were undergoing secretory expression, would trigger rapid cell death (**Figure** 7) if the presence of their preproteins had exceeded their individual allowable thresholds, the "Critical Values (CV)." A CV is defined as the largest quotient between an intracellular

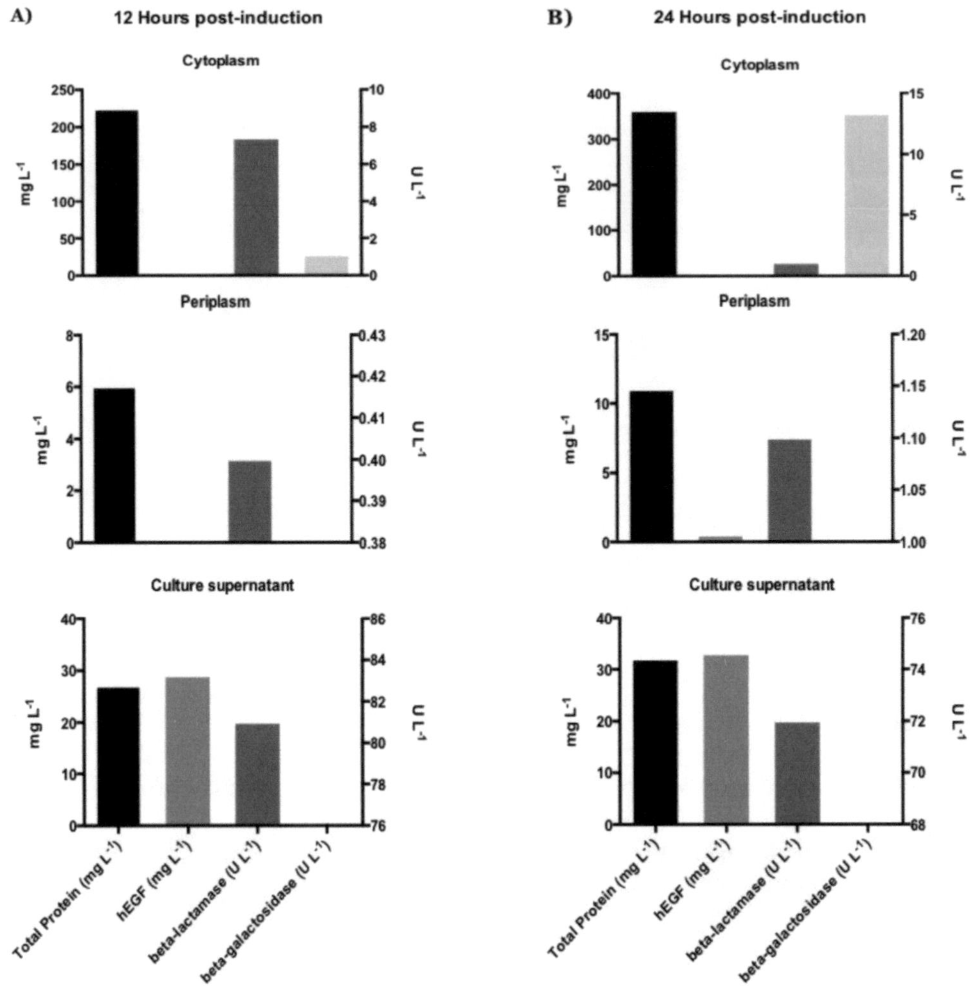

Figure 6.
Excretory production of EGF in E. coli. Protein concentrations and enzymatic activities detected in different subcellular compartments: cytoplasm, periplasm, and culture supernatant of E. coli transformants expressing EGF are shown. The samples prepared for various measurements were taken from the E. coli cell culture grown for 12 (A) and 24 h (B) after IPTG induction. The culture conditions employed were as described previously [80]. The results show that the EGF activities detected in the culture supernatant samples were the highest in all three compartments. However, beta-galactosidase activity was undetectable in the supernatant samples, supporting the conclusion that EGF activities detected in the supernatant samples resulted from excretion rather than from cell lysis.

preprotein and its secreted mature counterpart that was tolerable by the host cells [92]. Deletion of the signal peptide from its mature partner, despite the possibility of incurring the formation of inclusion bodies [48], interestingly it would help avoid the onset of the deadly effect resulting from an efficiently expressed secretory protein [110]. The results clearly indicate that the bottleneck of secretory/excretory production of a heterologous protein is at the stage of secretion.

Different approaches have been attempted to attain or even improve the CV, and hence the maximum production of a secretory heterologous protein on a per cell basis. Since cell death results from hyper-expression, strategies of optimizing, rather than maximizing, protein expression, e.g., less efficient promoters [92, 107, 111] and start codon [92], as well as defined minimal media and sub-optimal cultivation conditions [26] have been employed and shown to provide beneficial effects. More encouragingly, excretory production of Exg was

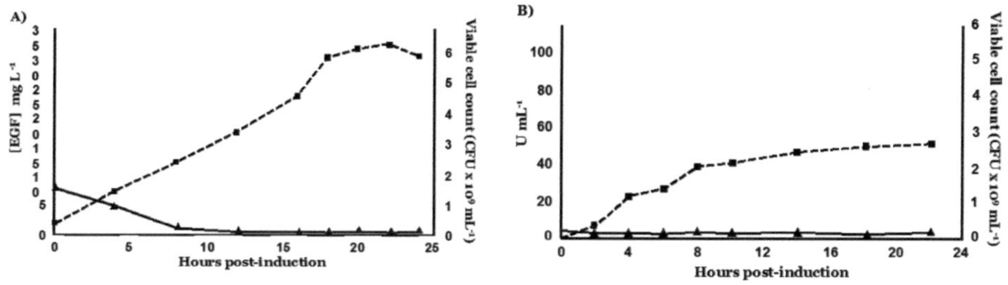

Figure 7.
Time-course experiments on extracellular expression of EGF and Exg in E. coli. (A) Measurements of EGF activities detected in the culture medium (–■–) and viable cell counts (–▲–) of E. coli transformants harboring plasmid pWKW2 are shown (unpublished results). (B) Measurements of Exg activities detected in the culture medium (–■–) and viable cell counts (–▲–) of E. coli transformants harboring plasmid tacIQpar8cex are shown. One unit of Exg activity in hydrolyzing p-nitrophenyl-β-D-cellobioside is defined as one nmol of p-nitrophenol produced per min. The growth conditions and IPTG induction of the cultures were done as described previously [111].

markedly enhanced when the level of Phage shock protein A (PspA) was elevated in the same host [109]. In the presence of additional PspA, the CV of secretory Exg was shown to be markedly increased from 20/80 to 45/55 [109]. Presumably, PspA helped the host cells to maintain membrane integrity and an energized membrane [112–114], which was readily equipped to cope with the "stress," the presence of secretory Pre-Exg, by efficiently transporting it through the SecYEG pathway [109].

5. Conclusions

Since the advent of recombinant DNA technology in the 1970s, *E. coli* has been the most favorable host choice for the expression of heterologous proteins. Strategies including both intracellular and secretory methods have been designed for the expression of proteins of interest. Despite possessing an outer membrane, a wide variety of naturally secreted proteins including hormones, factors, and degradative enzymes have also been shown to be produced as extracellular (excreted) products in *E. coli*. In undertaking both intracellular and secretory/excretory protein expression, a fusion approach is commonly adopted. Affinity tag proteins including β-galactosidase, glutathione *S*-transferase, and 6xHis-tag have been employed to form fusion precursors with desired proteins. To enable separation between the tag and target proteins, a protease cleavage site is required to be placed between the two proteins. However, on the one hand, it may be difficult to achieve the exact processing result through proteolytic cleavage. On the other hand, it is cost-ineffective to implement proteolytic cleavage on a large scale. Fusions of target proteins with inteins and secretion signal peptides have presented a practical approach to protein cleavages in cells without relying on the use of external proteases. It has been well demonstrated that using both methods of protein fusion, target proteins possessing the exactly processed sequences are obtainable through autocatalytic or signal peptidase cleavages in vivo in *E. coli*.

Acknowledgements

This study was funded by RGC project: GRF16101515 and Research Contracts: 13142580CLIL07W011 and 1617219-0.

Conflict of interest

The authors declare that they have no conflict of interest.

Abbreviations

aa	amino acid(s)
bFGF	basic fibroblast growth factor
β-gal	β-galactosidase
CBD	chitin-binding domain
C. fimi	*Cellulomonas fimi*
CNBr	cyanogen bromide
CV	critical value
DnaB	DnaB helicase
EGF	epidermal growth factor
Eng	endoglucanse
Exg	exoglucanase
E. coli	*Escherichia coli*
fMet	N-formyl-methionine
GST	glutathione S-transferase
I	Intein
IPTG	isopropyl β-D-1-thiogalactopyranoside
kDa	kilo dalton(s)
MBP	maltose-binding protein
Met	methionine
OmpA	outer membrane protein A
PhoA	alkaline phosphatase
Pre-	premature
PspA	phage shock protein A
PTM	post-translational modification
Sce	*Saccharomyces cerevisiae*
SDA	self-driven approach
SecYEG	trimeric complex of SecY, SecE, and SecG
Som	somatostatin
Ssp	*Synechocystis* species
TP	target protein
VMA	vacuolar ATPase subunit

Author details

Wan-Keung Raymond Wong*, Ka-Lun Ng, Thiyagarajan Sivakumar, Xiu-Hua Hu, Hao Wang and Lai-Cheuk Nelson Lai
Division of Life Science, The Hong Kong University of Science and Technology, Clear Water Bay, Kowloon, Hong Kong, China

*Address all correspondence to: bcwkrw@ust.hk

References

[1] Jackson DA, Symons RH, Berg P. Biochemical method for inserting new genetic information into DNA of Simian Virus 40: Circular SV40 DNA molecules containing lambda phage genes and the galactose operon of *Escherichia coli*. Proceedings of the National Academy of Sciences. 1972;**69**:2904-2909

[2] Cohen SN, Chang AC, Boyer HW, Helling RB. Construction of biologically functional bacterial plasmids in vitro. Proceedings of the National Academy of Sciences. 1973;**70**:3240-3244

[3] King G, Murray NE. Restriction enzymes in cells, not eppendorfs. Trends in Microbiology. 1994;**2**:465-469

[4] Matic I, Taddei F, Radman M. Genetic barriers among bacteria. Trends in Microbiology. 1996;**4**:69-73

[5] Kwong KW, Wong WKR. A revolutionary approach facilitating co-expression of authentic human epidermal growth factor and basic fibroblast growth factor in both cytoplasm and culture medium of *Escherichia coli*. Applied Microbiology and Biotechnology. 2013;**97**:9071-9080

[6] Kwong KW, Sivakumar T, Wong WKR. Intein mediated hyper-production of authentic human basic fibroblast growth factor in *Escherichia coli*. Scientific Reports. 2016;**6**:33948

[7] Böhlen P, Baird A, Esch F, Ling N, Gospodarowicz D. Isolation and partial molecular characterization of pituitary fibroblast growth factor. Proceedings of the National Academy of Sciences. 1984;**81**:5364-5368

[8] Ferrer-Miralles N, Domingo-Espín J, Corchero JL, Vázquez E, Villaverde A. Microbial factories for recombinant pharmaceuticals. Microbial Cell Factories. 2009;**8**:17

[9] Itakura K, Hirose T, Crea R, Riggs AD, Heyneker HL, Bolivar F, et al. Expression in *Escherichia coli* of a chemically synthesized gene for the hormone somatostatin. Science. 1977;**198**:1056-1063

[10] Goeddel DV, Kleid DG, Bolivar F, Heyneker HL, Yansura DG, Crea R, et al. Expression in *Escherichia coli* of chemically synthesized genes for human insulin. Proceedings of the National Academy of Sciences. 1979;**76**:106-110

[11] Goeddel DV, Heyneker HL, Hozumi T, Arentzen R, Itakura K, Yansura DG, et al. Direct expression in *Escherichia coli* of a DNA sequence coding for human growth hormone. Nature. 1979;**281**:544

[12] Goeddel DV, Shepard HM, Elizabeth Y, Leung D, Crea R, Sloma A, et al. Synthesis of human fibroblast interferon by *E. coli*. Nucleic Acids Research. 1980;**8**:4057-4074

[13] Goeddel DV, Yelverton E, Ullrich A, Heyneker HL, Miozzari G, Holmes W, et al. Human leukocyte interferon produced by *E. coli* is biologically active. Nature. 1980;**287**:411

[14] Yelverton E, Leung D, Week P, Gray PW, Goeddel DV. Bacterial synthesis of a novel human leukocyte interferon. Nucleic Acids Research. 1981;**9**:731-741

[15] DeLamarter J, Mermod J, Liang C, Eliason J, Thatcher D. Recombinant murine GM-CSF from *E. coli* has biological activity and is neutralized by a specific antiserum. The EMBO Journal. 1985;**4**:2575-2581

[16] Moks T, Abrahmsén L, Österlöf B, Josephson S, Östling M, Enfors S-O, et al. Large–scale affinity purification of human insulin-like growth factor I from culture medium of *Escherichia coli*. Nature Biotechnology. 1987;**5**:379

[17] Liu Q, Lin J, Liu M, Tao X, Wei D, Ma X, et al. Large scale preparation of recombinant human parathyroid hormone 1-84 from *Escherichia coli*. Protein Expression and Purification. 2007;**54**:212-219

[18] Nguyen AN, Song J-A, Nguyen MT, Do BH, Kwon GG, Park SS, et al. Prokaryotic soluble expression and purification of bioactive human fibroblast growth factor 21 using maltose-binding protein. Scientific Reports. 2017;**7**:16139

[19] Díaz M, Venturini E, Marchetti S, Arenas G, Marshall SH. Intein-mediated expression of cecropin in *Escherichia coli*. Electronic Journal of Biotechnology. 2012;**15**:3-3

[20] Sharma SS, Chong S, Harcum SW. Intein-mediated protein purification of fusion proteins expressed under high-cell density conditions in *E. coli*. Journal of Biotechnology. 2006;**125**:48-56

[21] Oka T, Sakamoto S, Miyoshi K, Fuwa T, Yoda K, Yamasaki M, et al. Synthesis and secretion of human epidermal growth factor by *Escherichia coli*. Proceedings of the National Academy of Sciences. 1985;**82**:7212-7216

[22] Dodt J, Schmitz T, Schäfer T, Bergmann C. Expression, secretion and processing of hirudin in *E. coli* using the alkaline phosphatase signal sequence. FEBS Letters. 1986;**202**:373-377

[23] Hsiung HM, Mayne NG, Becker GW. High-level expression, efficient secretion and folding of human growth hormone in *Escherichia coli*. Nature Biotechnology. 1986;**4**:991

[24] Greenberg R, Lundell D, Alroy Y, Bonitz S, Condon R, Fossetta J, et al. Expression of biologically active, mature human granulocyte-macrophage colony stimulating factor with an *E. coli* secretory expression system. Current Microbiology. 1988;**17**:321-332

[25] Huang R, Lam E, Chen Y, Hackett J, Lam T, Liu D, et al. Human epidermal growth factor excreted by recombinant *Escherichia coli* K-12 has the correct N-terminus and is fully bioactive. Process Biochemistry. 1999;**35**:1-5

[26] Sivakesava S, Xu Z, Chen Y, Hackett J, Huang R, Lam E, et al. Production of excreted human epidermal growth factor (hEGF) by an efficient recombinant *Escherichia coli* system. Process Biochemistry. 1999;**34**:893-900

[27] Willson C, Pebrin D, Cohn M, Jacob F, Monod J. Non-inducible mutants of the regulator gene in the "lactose" system of *Escherichia coli*. Journal of Molecular Biology. 1964;**8**:582-592

[28] Ullmann A, Jacob F, Monod J. Characterization by in vitro complementation of a peptide corresponding to an operator-proximal segment of the β-galactosidase structural gene of *Escherichia coli*. In: Selected Papers in Molecular Biology by Jacques Monod. Amsterdam: Elsevier; 1978. pp. 670-674

[29] Casadaban MJ, Martinez-Arias A, Shapira SK, Chou J. [21] β-Galactosidase gene fusions for analyzing gene expression in *Escherichia coli* and yeast. In: Methods in Enzymology. Amsterdam: Elsevier; 1983. pp. 293-308

[30] Brazeau P, Vale W, Burgus R, Ling N, Butcher M, Rivier J, et al. Hypothalamic polypeptide that inhibits the secretion of immunoreactive pituitary growth hormone. Science. 1973;**179**:77-79

[31] Hernan RA, Hui HL, Andracki ME, Noble RW, Sligar SG, Walder JA, et al. Human hemoglobin expression in *Escherichia coli*: Importance of optimal codon usage. Biochemistry. 1992;**31**:8619-8628

[32] Yasueda H, Nagase K, Hosoda A, Akiyama Y, Yamada K. High-level direct expression of semi-synthetic human interleukin-6 in *Escherichia coli* and production of N-terminus met-free product. Nature Biotechnology. 1990;**8**:1036

[33] Schellekens H, Casadevall N. Immunogenicity of recombinant human proteins: Causes and consequences. Journal of Neurology. 2004;**251**:ii4-ii9

[34] Liao YD, Jeng JC, Wang CF, Wang SC, Chang ST. Removal of N-terminal methionine from recombinant proteins by engineered *E. coli* methionine aminopeptidase. Protein Science. 2004;**13**:1802-1810

[35] Hirel P-H, Schmitter M, Dessen P, Fayat G, Blanquet S. Extent of N-terminal methionine excision from *Escherichia coli* proteins is governed by the side-chain length of the penultimate amino acid. Proceedings of the National Academy of Sciences. 1989;**86**:8247-8251

[36] Ben-Bassat A, Bauer K, Chang S-Y, Myambo K, Boosman A, Chang S. Processing of the initiation methionine from proteins: Properties of the *Escherichia coli* methionine aminopeptidase and its gene structure. Journal of Bacteriology. 1987;**169**:751-757

[37] Hwang DD, Li-Fan L, Lih-Yuan L. Co-expression of glutathione S-transferase with methionine aminopeptidase: A system of producing enriched N-terminal processed proteins in *Escherichia coli*. Biochemical Journal. 1999;**338**:335-342

[38] Piatkov KI, Vu TT, Hwang C-S, Varshavsky A. Formyl-methionine as a degradation signal at the N-termini of bacterial proteins. Microbial Cell. 2015;**2**:376

[39] Morino T, Morita M, Seya K, Sukenaga Y, Kato K, Nakamura T. Construction of a runaway vector and its use for a high-level expression of a cloned human superoxide dismutase gene. Applied Microbiology and Biotechnology. 1988;**28**:170-175

[40] Trepod C, Mott J. A spontaneous runaway vector for production-scale expression of bovine somatotropin from *Escherichia coli*. Applied Microbiology and Biotechnology. 2002;**58**:84-88

[41] Phillips M, Coffino P, Wang C. *Trypanosoma brucei* ornithine decarboxylase: Enzyme purification, characterization, and expression in *Escherichia coli*. The Journal of Biological Chemistry. 1988;**263**:17933-17941

[42] De Boer HA, Comstock LJ, Vasser M. The *tac* promoter: A functional hybrid derived from the *trp* and *lac* promoters. Proceedings of the National Academy of Sciences. 1983;**80**:21-25

[43] Wagner LA, Gesteland RF, Dayhuff TJ, Weiss RB. An efficient Shine-Dalgarno sequence but not translation is necessary for *lac*Z mRNA stability in *Escherichia coli*. Journal of Bacteriology. 1994;**176**:1683-1688

[44] Remaut E, Tsao H, Fiers W. Improved plasmid vectors with a thermoinducible expression and temperature-regulated runaway replication. Gene. 1983;**22**:103-113

[45] Lee JM, Lee J, Kim T, Lee SK. Switchable gene expression in *Escherichia coli* using a miniaturized photobioreactor. PLoS One. 2013;**8**:e52382

[46] Hale RS, Thompson G. Codon optimization of the gene encoding a domain from human type 1 neurofibromin protein results in a threefold improvement in expression level in *Escherichia coli*.

Protein Expression and Purification. 1998;**12**:185-188

[47] Wang L, Brock A, Herberich B, Schultz PG. Expanding the genetic code of *Escherichia coli*. Science. 2001;**292**:498-500

[48] O'Neill G, Kilburn D, Warren R, Miller R. Overproduction from a cellulase gene with a high guanosine-plus-cytosine content in *Escherichia coli*. Applied and Environmental Microbiology. 1986;**52**:737-743

[49] Wang Y, Slade MB, Gooley AA, Atwell BJ, Williams KL. Cellulose-binding modules from extracellular matrix proteins of *Dictyostelium discoideum* stalk and sheath. European Journal of Biochemistry. 2001;**268**:4334-4345

[50] Fox JD, Waugh DS. Maltose-binding protein as a solubility enhancer. In: *E. coli* Gene Expression Protocols. New York City: Springer; 2003. pp. 99-117

[51] Smith DB, Johnson KS. Single-step purification of polypeptides expressed in *Escherichia coli* as fusions with glutathione S-transferase. Gene. 1988;**67**:31-40

[52] Butt TR, Edavettal SC, Hall JP, MR M. SUMO fusion technology for difficult-to-express proteins. Protein Expression and Purification. 2005;**43**:1-9

[53] Kuo D, Nie M, Courey AJ. SUMO as a solubility tag and in vivo cleavage of SUMO fusion proteins with *Ulp1*. In: Protein Affinity Tags. New York City: Springer; 2014. pp. 71-80

[54] Arnau J, Lauritzen C, Petersen GE, Pedersen J. Current strategies for the use of affinity tags and tag removal for the purification of recombinant proteins. Protein Expression and Purification. 2006;**48**:1-13

[55] Wong WKR, Ng KL, Lam CC, Hu X, Lai N. Review article: Reasons for underrating the potential of human epidermal growth factor in medical applications. Journal of Analytical & Pharmaceutical Research. 2017;**4**:00101

[56] Fernández-Montequín JI, Betancourt BY, Leyva-Gonzalez G, Mola EL, Galán-Naranjo K, Ramírez-Navas M, et al. Intralesional administration of epidermal growth factor-based formulation (Heberprot-P) in chronic diabetic foot ulcer: Treatment up to complete wound closure. International Wound Journal. 2009;**6**:67-72

[57] Wong WKR. Effective treatment of an unhealed incision of a diabetic patient with recombinant human epidermal growth factor. Modern Chemistry & Applications. 2015;**3**:166

[58] Hirata R, Ohsumk Y, Nakano A, Kawasaki H, Suzuki K, Anraku Y. Molecular structure of a gene, *VMA1*, encoding the catalytic subunit of H(+)-translocating adenosine triphosphatase from vacuolar membranes of *Saccharomyces cerevisiae*. The Journal of Biological Chemistry. 1990;**265**:6726-6733

[59] Shah NH, Muir TW. Inteins: Nature's gift to protein chemists. Chemical Science. 2014;**5**:446-461

[60] Kwong KW, Ng AL, Wong WKR. Engineering versatile protein expression systems mediated by inteins in *Escherichia coli*. Applied Microbiology and Biotechnology. 2016;**100**:255-262

[61] Wong WKR. Means and method for expression of authentic human epidermal growth factor and/or basic fibrolast growth factor in cytoplasm and/or culture medium of *Escherichia coli*. 2017; US Patent No. 9,580,484 B2

[62] Wong WKR. Means and method for expression of authentic human

epidermal growth factor and/or basic fibrolast growth factor in cytoplasm and/or culture medium of *Escherichia coli*. 2017; Hong Kong Short-Term Patent No. HK1219200

[63] Wong WKR, Kwong WY. Improved means and method for hyper-production of authentic human basic fibroblast growth factor in *Escherichia coli*. 2017; Hong Kong Short-Term Patent No. HK1228651

[64] Sun Z, Chen J, Yao H, Liu L, Wang J, Zhang J, et al. Use of *Ssp* dnaB derived mini-intein as a fusion partner for production of recombinant human brain natriuretic peptide in *Escherichia coli*. Protein Expression and Purification. 2005;**43**:26-32

[65] Guo C, Li Z, Shi Y, Xu M, Wise JG, Trommer WE, et al. Intein-mediated fusion expression, high efficient refolding, and one-step purification of gelonin toxin. Protein Expression and Purification. 2004;**37**:361-367

[66] Hackenberger CP, Chen MM, Imperiali B. Expression of N-terminal Cys-protein fragments using an intein refolding strategy. Bioorganic & Medicinal Chemistry. 2006;**14**:5043-5048

[67] Zhang Y, Zhang K, Wan Y, Zi J, Wang Y, Wang J, et al. A pH-induced, intein-mediated expression and purification of recombinant human epidermal growth factor in *Escherichia coli*. Biotechnology Progress. 2015;**31**:758-764

[68] Chong S, Shao Y, Paulus H, Benner J, Perler FB, Xu M-Q. Protein splicing involving the *Saccharomyces cerevisiae* VMA intein. The steps in the splicing pathway, side reactions leading to protein cleavage, and establishment of an in vitro splicing system. The Journal of Biological Chemistry. 1996;**271**:22159-22168

[69] Villa-Komaroff L, Efstratiadis A, Broome S, Lomedico P, Tizard R, Naber SP, et al. A bacterial clone synthesizing proinsulin. Proceedings of the National Academy of Sciences. 1978;**75**:3727-3731

[70] Talmadge K, Stahl S, Gilbert W. Eukaryotic signal sequence transports insulin antigen in *Escherichia coli*. Proceedings of the National Academy of Sciences. 1980;**77**:3369-3373

[71] Talmadge K, Kaufman J, Gilbert W. Bacteria mature preproinsulin to proinsulin. Proceedings of the National Academy of Sciences. 1980;**77**:3988-3992

[72] Talmadge K, Gilbert W. Cellular location affects protein stability in *Escherichia coli*. Proceedings of the National Academy of Sciences. 1982;**79**:1830-1833

[73] Miyake T, Oka T, Nishizawa T, Misoka F, Fuwa T, Yoda K, et al. Secretion of human interferon-α induced by using secretion vectors containing a promoter and signal sequence of alkaline phosphatase gene of *Escherichia coli*. The Journal of Biochemistry. 1985;**97**:1429-1436

[74] Lopes J, Gottfried S, Rothfield L. Leakage of periplasmic enzymes by mutants of *Escherichia coli* and *Salmonella typhimurium*: isolation of "periplasmic leaky" mutants. Journal of Bacteriology. 1972;**109**:520-525

[75] Singh A, Reithmeier R. Leakage of periplasmic enzymes from cells of heptose-deficient mutants of *Escherichia coli*, associated with alterations in the protein component of the outer membrane. The Journal of General and Applied Microbiology. 1975;**21**:109-118

[76] Collmer A, Wilson DB. Cloning and expression of a Thermomonospora

YX endocellulase gene in *E. coli*. Nature Biotechnology. 1983;**1**:594

[77] Gilkes N, Kilburn D, Langsford M, Miller R Jr, Wakarchuk W, Warren R, et al. Isolation and characterization of *Escherichia coli* clones expressing cellulase genes from *Cellulomonas fimi*. Microbiology. 1984;**130**:1377-1384

[78] Kudo T, Kato C, Horikoshi K. Excretion of the penicillinase of an alkalophilic *Bacillus* sp. through the *Escherichia coli* outer membrane. Journal of Bacteriology. 1983;**156**:949-951

[79] Nagahari K, Kanaya S, Munakata K, Aoyagi Y, Mizushima S. Secretion into the culture medium of a foreign gene product from *Escherichia coli*: use of the ompF gene for secretion of human beta-endorphin. The EMBO Journal. 1985;**4**:3589-3592

[80] Wong WKR, Sutherland ML. Excretion of heterologous proteins from *E. coli*. 1993; US Patent No. 5223407

[81] Ito K, Beckwith JR. Role of the mature protein sequence of maltose-binding protein in its secretion across the *E. coli* cytoplasmic membrane. Cell. 1981;**25**:143-150

[82] Bieker KL, Silhavy TJ. PrlA is important for the translocation of exported proteins across the cytoplasmic membrane of *Escherichia coli*. Proceedings of the National Academy of Sciences. 1989;**86**: 968-972

[83] Bedouelle H, Bassford PJ Jr, Fowler AV, Zabin I, Beckwith J, Hofnung M. Mutations which alter the function of the signal sequence of the maltose binding protein of *Escherichia coli*. Nature. 1980;**285**:78

[84] Lee C, Li P, Inouye H, Brickman E, Beckwith J. Genetic studies on the inability of beta-galactosidase to be translocated across the *Escherichia coli* cytoplasmic membrane. Journal of Bacteriology. 1989;**171**:4609-4616

[85] Huber D, Boyd D, Xia Y, Olma MH, Gerstein M, Beckwith J. Use of thioredoxin as a reporter to identify a subset of *Escherichia coli* signal sequences that promote signal recognition particle-dependent translocation. Journal of Bacteriology. 2005;**187**:2983-2991

[86] Chatzi KE, Sardis MF, Economou A, Karamanou S. SecA-mediated targeting and translocation of secretory proteins. Biochimica et Biophysica Acta (BBA)—Molecular Cell Research. 2014;**1843**:1466-1474

[87] Park S, Schumann W. Optimization of the secretion pathway for heterologous proteins in *Bacillus subtilis*. Biotechnology and Bioprocess Engineering. 2015;**20**:623-633

[88] Tsirigotaki A, De Geyter J, Šoštaric N, Economou A, Karamanou S. Protein export through the bacterial Sec pathway. Nature Reviews. Microbiology. 2017;**15**:21

[89] Freudl R. Signal peptides for recombinant protein secretion in bacterial expression systems. Microbial Cell Factories. 2018;**17**:52

[90] Furukawa A, Nakayama S, Yoshikaie K, Tanaka Y, Tsukazaki T. Remote coupled drastic β-barrel to β-sheet transition of the protein translocation motor. Structure. 2018;**26**:485.e2-489.e2

[91] Wong WKR, Gerhard B, Guo ZM, Kilburn DG, Anthony R, Warren J, et al. Characterization and structure of an endoglucanase gene cenA of *Cellulomonas fimi*. Gene. 1986;**44**:315-324

[92] Fu ZB, Ng KL, Lam CC, Leung KC, Yip WH, Wong WKR. A

two-stage refinement approach for the enhancement of excretory production of an exoglucanase from *Escherichia coli*. Protein Expression and Purification. 2006;**48**:205-214

[93] Wülfing C, Plückthun A. Protein folding in the periplasm of *Escherichia coli*. Molecular Microbiology. 1994;**12**:685-692

[94] Salinas G, Pellizza L, Margenat M, Flo M, Fernandez C. Tuned *Escherichia coli* as a host for the expression of disulfide-rich proteins. Biotechnology Journal. 2011;**6**:686-699

[95] Wong WKR, Fu ZB, Wang Y, Ng KL, Chan KN. Engineering of efficient *Escherichia coli* excretion systems for the production of heterologous proteins for commercial applications. Recent Patents on Chemical Engineering. 2012;**5**:45-55

[96] Gilkes NR, Langsford ML, Kilburn DG, Miller R, Warren R. Mode of action and substrate specificities of cellulases from cloned bacterial genes. The Journal of Biological Chemistry. 1984;**259**:10455-10459

[97] Remaut E, Stanssens P, Fiers W. Plasmid vectors for high-efficiency expression controlled by the PL promoter of coliphage lambda. Gene. 1981;**15**:81-93

[98] Studier FW, Moffatt BA. Use of bacteriophage T7 RNA polymerase to direct selective high-level expression of cloned genes. Journal of Molecular Biology. 1986;**189**:113-130

[99] Curry K, Tomich C-S. Effect of ribosome binding site on gene expression in *Escherichia coli*. DNA. 1988;**7**:173-179

[100] Movva N, Nakamura K, Inouye M. Amino acid sequence of the signal peptide of ompA protein, a major outer membrane protein of *Escherichia coli*. The Journal of Biological Chemistry. 1980;**255**:27-29

[101] Calos MP. DNA sequence for a low-level promoter of the *lac* repressor gene and an 'up'promoter mutation. Nature. 1978;**274**:762

[102] Lanzer M, Bujard H. Promoters largely determine the efficiency of repressor action. Proceedings of the National Academy of Sciences. 1988;**85**:8973-8977

[103] Minton NP. Improved plasmid vectors for the isolation of translational lac gene fusions. Gene. 1984;**31**:269-273

[104] Wong DK, Lam K, Chan C, Wong Y, Wong WKR, Hackett J. Extracellular expression of human epidermal growth factor encoded by an *Escherichia coli* K-12 plasmid stabilized by the *ytl2-incR* system of *Salmonella typhimurium*. Journal of Industrial Microbiology & Biotechnology. 1998;**21**:31-36

[105] Wong WKR, RC Huang EL, Wong RS, Morris C, Hackett J. Applications, and efficient large-scale production, of recombinant human epidermal growth factor. Biotechnology & Genetic Engineering Reviews. 2001;**18**:51-71

[106] Guo Z, Arfman N, Ong E, Gilkes N, Kilburn D, Warren R, et al. Leakage of *Cellulomonas fimi* cellulases from *Escherichia coli*. FEMS Microbiology Letters. 1988;**49**:279-283

[107] Wang YY, Ng KL, Lam CC, Chan A, Sze K, Fu Z, et al. Efficient *Bacillus subtilis* promoters for graded expression of heterologous genes in *Escherichia coli*. Research Journal of Biotechnology. 2010;**5**:5-14

[108] O'Neill G, Goh SH, Warren RA, Kilburn DG, Miller RC Jr. Structure of the gene encoding the exoglucanase of *Cellulomonas fimi*. Gene. 1986;**44**:325-330

[109] Wang YY, Fu Z, Ng KL, Lam CC, Chan A, Sze K, et al. Enhancement of excretory production of an exoglucanase from *Escherichia coli* with phage shock protein A (PspA) overexpression. Journal of Microbiology and Biotechnology. 2011;**21**:637-645

[110] Fu ZB, Ng KL, Lam TL, Wong WKR. Cell death caused by hyper-expression of a secretory exoglucanase in *Escherichia coli*. Protein Expression and Purification. 2005;**42**:67-77

[111] Lam TL, Wong RS, Wong WKR. Enhancement of extracellular production of a *Cellulomonas fimi* exoglucanase in *Escherichia coli* by the reduction of promoter strength. Enzyme and Microbial Technology. 1997;**20**:482-488

[112] Kleerebezem M, Crielaard W, Tommassen J. Involvement of stress protein PspA (phage shock protein A) of *Escherichia coli* in maintenance of the protonmotive force under stress conditions. The EMBO Journal. 1996;**15**:162-171

[113] DeLisa MP, Lee P, Palmer T, Georgiou G. Phage shock protein PspA of *Escherichia coli* relieves saturation of protein export via the Tat pathway. Journal of Bacteriology. 2004;**186**:366-373

[114] Male A, Tavassoli A. Self-assembly of *Escherichia coli* phage shock protein A. Advances in Microbiology. 2014;**4**:353-359

Chapter 3

Enteropathogenic *Escherichia coli*

Isabel Cristina Affonso Scaletsky

Abstract

The term enteropathogenic *Escherichia coli* (EPEC) was first used in 1955 to describe a number of serogroup-defined *E. coli* strains associated with infantile diarrhea. EPEC are now defined as those that produce a characteristic intestinal histopathology known as attaching and effacing (A/E) and do not produce Shiga toxins. EPEC carry the locus for enterocyte effacement (LEE) pathogenicity island, which contains the *eae* gene that encodes an outer membrane protein called intimin. Typical EPEC (tEPEC) carry a virulence plasmid known as the pEAF (EPEC adhesion factor plasmid) which encodes the bundle-forming pilus (BFP) that mediate localized adherence to epithelial cells, whereas atypical EPEC (aEPEC) do not possess this plasmid. Typical EPEC strains have been associated with severe outbreaks of infant diarrhea in developing countries. Atypical EPEC strains have been linked to diarrhea outbreaks at all ages worldwide. Diarrhea due to aEPEC in children is not as severe as that caused by tEPEC.

Keywords: *Escherichia coli*, EPEC, diarrhea, children, infantile diarrhea

1. History and definition of EPEC

Escherichia coli were first recognized as diarrheal pathogens in 1898, when Lesage demonstrated that serum from diarrhea patients agglutinated strains of *E. coli* isolated from other patients in the same outbreak but not those of control [1]. In 1945, Bray discovered that *E. coli* strains of certain serogroups were the predominant cause of summer diarrhea in infants in the United Kingdom [2]. In 1947, Kauffman published a serotyping scheme based on somatic (O), flagellar (H), and capsular (K) antigens, providing a reliable method of typing diarrheagenic *E. coli* [3].

The term enteropathogenic *Escherichia coli* (EPEC) was introduced in 1955 to describe strains of *E. coli* implicated epidemiologically with infant diarrhea in the 1940s and 1950s [4]. This definition changed as additional serotypes were associated with infantile diarrhea. EPEC were recognized as important causes of infant diarrhea in the 1950s and 1960s in the developed world and subsequently have been shown to be common agents of gastroenteritis in the developing world. The confirmation that EPEC strains were pathogenic came from human volunteer studies carried out by Levine et al. [5].

The definition of EPEC has changed as various mechanisms of pathogenesis have been discovered. During the late 1960s and early 1970s, two other diarrheagenic *E. coli* strains were discovered, strains producing the heat-stable enterotoxin (ST) and the heat-labile enterotoxin (LT) were designated enterotoxigenic *E. coli* (ETEC), and strains demonstrating *Shigella*-like invasiveness were designated enteroinvasive *E. coli* (EIEC). At this time, the original definition of EPEC by Neter [4] has undergone modification considerably. EPEC were then defined as "diarrheagenic *E. coli* belonging to serogroups epidemiologically incriminated as pathogens but whose pathogenic mechanisms have not been proven to be related either heat-labile (LT) or heat-stable enterotoxins (ST) or to *Shigella*-like invasiveness" [6].

In 1979, the first phenotype characteristic other than serotyping associated with EPEC was the observation of Cravioto et al. [7] that 80% of EPEC strains as defined by serotype could adhere to HEp-2 cells in cell culture, while most non-EPEC strains could not. Later, the adherence pattern of EPEC was described as "localized adherence" (LA), based on the presence of clusters or microcolonies on the surface of HEp-2 cells [8]. Baldini et al. [9] subsequently showed that the ability of EPEC E2348/69 strain (O127:H6) to adhere in a localized adherence pattern was associated with the presence of a 60-MDa plasmid. EPEC strains representing a variety of serotypes were found to possess highly conserved high molecular weight plasmids associated with localized adherence, the so-called EPEC adherence factor (EAF) plasmids [10, 11]. Subsequently, differences in adherence patterns were discerned by Scaletsky et al. [8] and Nataro et al. [12], giving rise to two other categories of diarrheagenic *E. coli* and diffusely adherent *E. coli* and enteroaggregative *E. coli*. Also in the 1980s, a newly recognized clinical syndrome caused by *E. coli* led to the discovery that some diarrheagenic *E. coli* produce a potent cytotoxin known as Shiga toxin (Stx).

In 1983, Moon et al. [13] published electron micrographs of pigs and rabbits infected with EPEC and coined the term "attaching and effacing" (A/E) to describe the loss of microvilli, intimate attachment of the bacteria to the host, and formation of pedestals at the sits of bacteria attachment. In 1987, a number of studies clarified the relationship between LA phenotype and A/E lesion, which confirmed earlier reports that LA is associated with the EAF plasmid, and demonstrated for the first time that A/E is encoded on the chromosome [14].

Originally defined by serotype, EPEC are now defined as those having the ability to cause diarrhea, the ability to produce A/E histopathology on the intestinal epithelium, and the inability to produce Shiga toxins based on pathogenic characteristics [15]. Improvements in techniques allowing a better understanding of the genome and virulence mechanisms among EPEC strains over the years have led to the classification into "typical" and "atypical" subtypes based on the presence or absence of the pEAF plasmid [15].

2. Atypical versus typical EPEC

Most of the typical EPEC strains belong to the traditional EPEC serogroups O55, O86, O111, O114, O119, O127, and O142, and the most common flagellar antigens are H6 and H2 [16, 17]. A less common EPEC type is H34, and a number of typical EPEC strains are nonmotile in conventional testes and classified as H-. Typical EPEC strains belonging to nonclassic serotypes have also been reported [18, 19]. Currently, more than 180 different O serogroups and more than 60 H antigens are recognized. Atypical EPEC belong to a large diversity of classical and nonclassical serotypes [18, 20]. Based on multilocus enzyme electrophoresis analysis of allelic differences among housekeeping genes, typical EPEC strains have been subtyped into two major lineages, previously designated EPEC1 and EPEC2 [16, 17]. EPEC1 includes serotypes O55:H6 and O119:H6, whereas EPEC2 consists of serotypes O111:H2 and O114:H2. Recently, EPEC strains have been demonstrated to cluster in three main lineages, designated EPEC1, EPEC2, and EPEC4, which probably acquired the locus of enterocyte effacement region (LEE) and pEAF independently [21].

Interestingly, it has been found that 35% of the atypical EPEC strains also belong to the typical EPEC lineages [21]. Thus, it has been hypothesized that at least some atypical EPEC may have originated from typical EPEC strains that lost pEAF in the host or in the environment [21].

3. Epidemiology of EPEC

3.1 Incidence

The prevalence of EPEC infection varies between epidemiological studies based on differences in study populations, age, distributions, and methods used for detection and diagnosis [22]. Also, geographic region and socioeconomic class may contribute to the epidemiology of EPEC-induced diarrheal disease [23]. Adults and older children with typical EPEC infections are rarely reported; this has been attributed to the loss of specific receptors with age or development of immunity [24].

For the two last decades, studies conducted worldwide have shown the association of typical EPEC serotypes with diarrhea in children <1 year of age, mainly in poor children of urban centers [24]. This association was particularly strong in infants less than 6 months of age. Between 1977 and 1982, epidemiologic studies in Brazil, Chile, Mexico, and South Africa have shown that 30–40% of infantile diarrhea was caused by typical EPEC serotypes [22]. However, recent studies in these countries have not identified a significant association between typical EPEC and infantile diarrhea. At this time, a change in the epidemiology of EPEC occurred in both developing and developed countries. The proportion of atypical EPEC strains has increased and outnumbered typical EPEC strains, and atypical EPEC strains have also been associated with childhood diarrhea in both developing and developed countries [19, 22, 25]. In Brazil, 92% of EPEC isolates collected from children between 2001 and 2002 were atypical [26], compared to 38% in a 1998–1999 study [27–29]. However, other studies still report typical being more prevalent than atypical EPEC as a cause of diarrhea [30]. Recently, a prospective, population-based case-control study involving seven sites in Africa and Asia showed that typical EPEC was significantly associated with moderate to severe diarrhea in children under 2 years of age in Kenya, whereas atypical EPEC was not associated with this type of diarrhea [31].

3.2 Transmission and reservoirs

Typical EPEC transmission follows a fecal-oral process through contaminated surfaces, weaning fluids, and human carriers [32]. EPEC outbreaks among adults, although rare, seem to occur through ingestion of contaminated food and water; however, no specific environmental reservoir has been identified [24]. EPEC outbreaks have been reported to show a seasonal distribution with peaks during the warm months [33]. Humans are the only known reservoir for typical EPEC, with symptomatic and asymptomatic children and asymptomatic adults being the most likely source [24].

In contrast to the tEPEC, many aEPEC strains have been found in diarrheic as well as in healthy animals and from the environment. Interestingly, animal aEPEC serogroups associated with human diarrhea have been identified (e.g., O26, O103, O119, O128, O142, and O157); however, so far a direct transmission from animals to humans has not been confirmed. In addition, foods including raw meats, pasteurized milk, meat samples, vegetables, and water have been also implicated as vehicles of aEPEC to human infections (reviewed in [34, 35]).

4. EPEC virulence factors and genetics

4.1 Localized adherence

Typical EPEC strains adhere to HeLa, HEp-2, and other cell lines and to organ cultures in vitro in a distinctive pattern of three-dimensional microcolonies so-called localized adherence (LA) pattern within 3 h of infection (**Figure 1A**) [8, 24]. A similar adherence pattern has been seen in tissue biopsies of EPEC-infected humans [37]. The LA phenotype is mediated by a type IV fimbriae bundle-forming pilus (BFP) associated with the EAF plasmid, which mediates bacterium-to-bacterium adherence, resulting in formation of compact microcolonies [38].

Atypical EPEC strains may display a variant LA pattern designated LA-like (LAL) pattern, which is characterized by the presence of loose compact microcolonies or clusters of bacteria in few cells observed in tests using prolonged incubation periods (6 h) (**Figure 1B**) [39, 40]. Interestingly, the LAL pattern is determined in prolonged assays (6 h) of bacteria-cell interaction [39]. LAL is the most common pattern seen among EPEC strains; however, some strains display alternate adherence phenotypes such as diffuse adherence (DA) and aggregative adherence (AA) [19].

4.2 Attaching and effacing (A/E) lesion

The hallmark of EPEC infection is the ability of the organism to attach intimately to epithelial cells and efface microvilli (**Figure 1C**). This effect was first described by Staley et al. [41], although the term attaching and effacing (A/E)

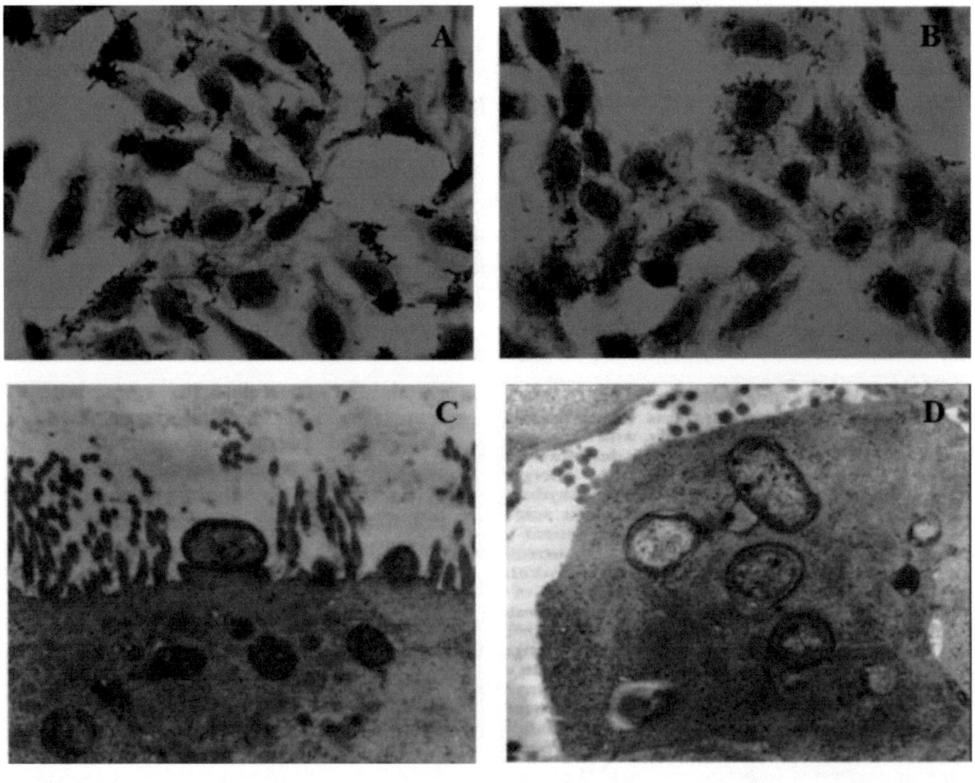

Figure 1.
EPEC adhesion to epithelial cells: (A) localized adherence pattern (LA) of typical EPEC on HeLa cells; (B) localized adherence-like (LAL) pattern of atypical EPEC on HeLa cells; (C) attaching and effacing (A/E) of enterocytes by EPEC; and (D) small bowel biopsy of infant infected with typical EPEC O111ab:H2 [36].

was coined by Moon et al. [13]. The A/E characteristic can be observed by electron microscopic examination of cultured epithelial cells exposed to EPEC or of intestinal biopsies from infants or animals infected with EPEC [24].

4.3 Invasiveness

Intracellular typical EPEC have been observed both in tissue culture and in small intestinal biopsies from an EPEC-infected infant (**Figure 1D**) [36]. Fletcher et al. [42] and Scaletsky et al. [43] have reported that EPEC O111:NM strains contain plasmid sequences that confer invasiveness upon *E. coli* K12 strains. However, despite their invasive potential in vitro, most EPEC are considered as noninvasive pathogens [44].

4.4 Biofilm formation

Typical EPEC have the ability to form biofilms on abiotic surfaces under static conditions, or on a flow through continuous culture system, and a model of EPEC biofilm formation has been proposed [45]. Biofilm formation requires adhesive structures as type 1 pili, antigen 43, BFP, and the EspA filament (see below) as participants in bacterial aggregation during biofilm formation on abiotic surfaces [45]. Atypical EPEC strains have also been shown to adhere to abiotic surfaces (polystyrene and glass) [46, 47]. The non-fimbrial adhesin curli and the T1P were shown to mediate binding to these surfaces in some atypical EPEC at different temperatures [48, 49].

4.5 The EAF plasmid

Typical EPEC strains possess a large virulence plasmid called the EPEC adherence factor (EAF) plasmid [9], which varies in sequence among different EPEC strains but is somewhat conserved [12]. The EAF plasmid pMAR2 is found among strains of the EPEC1 lineage, whereas pB171 is more common among EPEC2 strains [50, 51]. Two sets of genes located on the EAF plasmid are important for pathogenicity: the *bfp* gene cluster encoding BFP [38] and the *per* locus encoding a transcriptional activator called plasmid-encoded regulator (Per) [51]. Both BFP and PerA have been shown to contribute to virulence in human volunteers [52]. Between pMAR2 and pB171, the *bfp* and *per* loci share 99% sequence similarity [50]. Studies of comparison genomics of the EAF plasmids from varied EPEC phylogenomic lineages demonstrated significant plasmid diversity among isolates within the same phylogenomic lineage [53].

4.6 Bundle-forming pilus (BFP)

Typical EPEC strains produce a type IV pilus, the bundle-forming pili (BFP), which interconnects bacteria within microcolonies, promoting their stabilization and producing the LA phenotype [38]. The BFP is encoded by an operon of 14 genes contained on the EAF plasmid, with *bfpA* encoding the major structural subunit (bundlin) [54]. These 14 *bfp* genes are highly conserved among EPEC1 and EPEC2 strains. Some O128:H2 and O119:H2 EPEC strains that contain part of the *bfpA* gene have the rest of the *bfp* gene cluster deleted and replaced with an IS66 element [55, 56].

4.7 The locus of enterocyte effacement (LEE) and the type III secretion system (TTSS)

The locus of enterocyte effacement (LEE) is a 35.6-kb pathogenicity island of EPEC containing genes necessary for the formation of the A/E lesion [57]. The EPEC LEE contains at least 41 open reading frames that are organized into five operons

(*LEE1* to *LEE5*) [58–60]. *LEE1*, *LEE2*, and *LEE3* encode a type III protein secretion system (T3SS) and Ler (LEE-encoded regulator) regulators, such as GrlA (global regulator of LEE activator, formerly called Orf11) and GrlR (Grl repressor, formerly called Orf10) [61]. *LEE4* encodes the EPEC-secreted proteins EspA, EspB, and EspD via the type III system. LEE5 encodes intimin and its translocated receptor, Tir [62]. Besides Tir, the EPEC genome contains other six LEE-encoded effector proteins translocated into the cell (Map, EspF, EspG, EspZ, EspH, and EspB), which interfere with different aspects of the cell physiology ([58, 59]; reviewed in [44]) [63].

In addition to the LEE effectors, various non-LEE (Nle)-encoded effector genes (*cif*, *espI/nleA*, *nleB*, *nleC*, *nleD*, *nleE*, and *nleH*) [59, 63] were described, which are located outside the LEE region of EPEC, in at least six chromosomal PAIs, or in prophage elements (reviewed in [64] and [65]). Although they are not required for AE lesion formation, it is understood that they contribute to increased bacterial virulence [66].

The LEE region of some atypical EPEC strains display a genetic organization similar to that found in the typical EPEC prototype E2348/69 strain [66]. Although the T3SS-encoding genes are considerably conserved [66, 67], the effector protein-encoding genes display important differences, and remarkable differences can be detected at the 5′ and 3′ flanking regions of atypical EPEC, suggesting the occurrence of different evolution events [68].

The expression of LEE genes is controlled by Per, which is encoded on the EAF plasmid present in typical EPEC strains. Per activates Ler, which in turn activates the *LEE2*, *LEE3*, *LEE4*, and *LEE5* operons, and the genes *espF*, *espG*, and *map* [58, 59]. The Ler protein is a histone-like nucleoid-structuring protein (H-NS) that responds to an environmental stimulus (temperature). Ler also controls genes located outside the LEE, such as *espC* and *nleA* [60]. Additional regulatory system has been shown to control expression of the LEE [69]. The AI-2 (autoinducer-2) quorum-sensing system regulates LEE1 operon, which increases expression of the *LEE3* and *LEE4* operons via the *ler* gene product. Two novel LEE-encoded regulators that have roles in *ler* expression were reported, GrlA (global regulator of LEE activator) and GrlR (Grl repressor) [61]. GrlR and GrlA are positive and negative regulators, respectively, required for the expression of several LEE-encoded genes [61]. Other LEE regulators include the integration host factor (IHF); Bip, a tyrosine-phosphorylated GTPase; Fis (factor for inversion stimulation); and GadX, which is a member of the AraC transcription factor family [58].

4.8 Intimin and Tir

Intimin is a 94-KDa outer membrane adhesin encoded by the *eae* gene and required for intimate adherence of EPEC to epithelial cells at the sites of A/E lesions [24]. N-Terminal portions are highly conserved, whereas C-terminus portions are highly variable [70]. C-Terminal intimin differences have been used as a basis for classification into several distinct subtypes (represented by the Greek letters to α (alpha) through ζ (zeta) [71, 72]); human EPEC1 strains express subtype α, while EPEC2 strains express subtype β. The N-terminus portion binds intimin in the bacterial outer membrane, whereas the C-terminus portion binds intimin to Tir. The binding of intimin to Tir leads to intimate adherence of the bacterium to the epithelium and pedestal formation beneath adherent bacteria. In addition, Tir inhibits NF-κB activity by targeting tumor necrosis factor alpha (TNF-α) receptor-associated factors [73].

4.9 Other potential adhesins

In addition to BFP, two other EPEC surface structures, rodlike fimbriae and fibrillae, have been characterized and have been suggested to be involved in the interaction of EPEC with host cells [74, 75]. Additionally, EPEC strains encode a large surface protein, lymphocyte inhibitory factor (LifA), that contributes to epithelial cell adherence in vitro [76, 77] and is required for intestinal colonization of mice by the related A/E pathogen *C. rodentium* [78]. The *lifA* gene is more commonly found among typical rather than atypical EPEC strains [79]; however, atypical EPEC strains harboring *lifA* have a significant association with diarrhea in children under 5 years of age [80]. A novel gene cluster, designated the locus for diffuse adherence (*lda*), was found in an atypical EPEC O26 strain that is responsible for mediating DA adherence; its expression is induced by bile salts [81]. The *E. coli* common pilus (ECP) has also been shown to act as an accessory adherence factor in EPEC, playing a role during cell adherence and/or in bacterium-bacterium interactions [82].

4.10 Flagella

Flagella has been suggested to be involved in EPEC adherence to epithelial cells [83]. EPEC mutants with transposon insertion in the flagellar gene *fliC* were deficient in localized adherence, and anti-flagellum antibodies were effective in blocking the adherence of several EPEC serotypes [83]. However, a subsequent study has not confirmed a role of flagella in EPEC adherence [84].

5. EspC

EspC is a high-molecular-weight secreted protein of EPEC that induces cytopathic effects on epithelial cells, including cytoskeletal damage [24, 85]. EspC is a member of the serine protease autotransporters of the *Enterobacteriaceae* (SPATE) family of autotransporter proteins that encodes its own transport mechanism. Moreover, espC has been shown to interact with and degrade hemoglobin [86] and to hydrolyze other proteins such as pepsin, factor V, and spectrin [87]. In addition, EspC confers enhanced lysozyme resistance to EPEC [87] and serves as a substratum for adherence and biofilm formation as well as to protect bacteria from antimicrobial compounds [88]. EspC is encoded in a 15-kb chromosomal island specific to EPEC1 strains [24].

6. Other toxins

Scott and Kaper [89] reported a cytolethal distending toxin (CDT) in an EPEC strain that induces chromatin disruption, which leads to G2/M-phase growth arrest of the target cell and ultimately cell death [90]. A study has suggested that most EPEC strains from diarrhea harbor the CDT gene [91]. Another toxin is the enteroaggregative *E. coli* heat-stable enterotoxin 1 (EAST1) also present in EPEC strains [92]. The use of an EAST1 DNA probe suggests that this toxin is expressed by a number of clinical EPEC isolates [18, 93]. The role of CDT and EAST1 in EPEC pathogenesis remains to be elucidated.

7. Model of EPEC pathogenesis

A three-stage model of EPEC pathogenesis was first described in the early 1990s [94], Clarke et al. [95], including localized adherence to the host cell, signal transduction, and intimate attachment with pedestal formation (**Figure 2**).

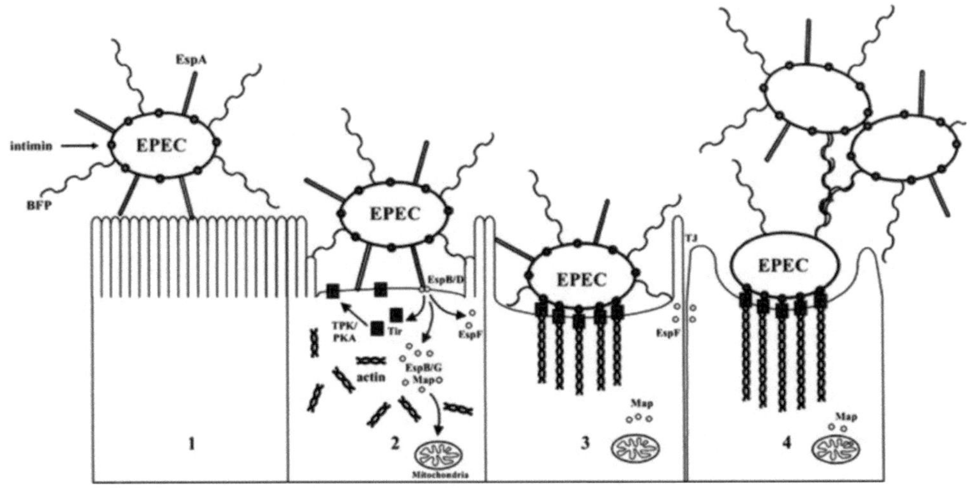

Figure 2.
Four-stage model of EPEC pathogenesis. Reprinted from Clarke et al. [95].

In the first stage, an attachment of typical EPEC to the surface of the host intestinal epithelium is mediated by the bundle-forming pili (BFP). The filament EspA also promotes attachment, albeit in a less efficient manner, and could mediate adherence of atypical EPEC strains. In the second stage, Tir and effector proteins (EspB, EspD, EspF, EspG, and Map), translocated into the host cells via type III system apparatus, activate cell-signaling pathways, causing alterations in the host cell cytoskeleton and resulting in actin accumulation and loss of microvilli [58]. In the third stage, bacteria intimately adhered to host cell by intimin-Tir interactions amplifies the accumulation of filaments of actin and other cytoskeletal proteins that result in pedestal-like structures [62, 96, 97]. Finally, the translocated effectors disrupt host cell processes, causing loss of tight-junction integrity and mitochondrial function and leading to both electrolyte loss and eventual cell death.

8. Diagnosis

Traditionally, the identification of EPEC was based on the O:H serotyping, but serotype designation is no longer precise. The identification of EPEC was based on the characteristic of EPEC's attachment to epithelial cells and may include phenotypic or genotypic tests. The HeLa adherence assay distinguishes EPEC from other *E. coli* by their ability to adhere in a localized pattern (LA) on the surface of cells [8]. The fluorescent actin-staining (FAS) assay, originally described by Knutton et al. [98], leads to the identification of the A/E lesion, by detecting actin condensation under EPEC adhesion pedestals. DNA probes and PCR targeting genes responsible for these characteristics were developed. A 1-kb EAF fragment probe was initially developed as a diagnostic DNA probe (the EAF probe) and subsequently refined as an oligonucleotide probe as well as PCR primers [10, 11, 99]. The identification of *bfpA*, the structural gene encoding BFP, led to the development of more specific and sensitive probe or PCR tests to detect typical EPEC strains [74, 75, 100]. However, some PCR primers may fail to identify all typical EPEC strains since multiple alleles of *bfpA* have been identified [101]. The *eae* sequences by DNA probes and PCR primers have been used to detect the presence of LEE encoding A/E lesion [24].

9. Clinical features of EPEC infection

9.1 Symptoms

The most common symptoms reported in EPEC infection are watery diarrhea, dehydration, vomiting, food intolerance, and low-grade fever [24, 97]. In addition, EPEC infection may lead to severe malabsorption of nutrients resulting in nutritional aggravation and persistence of diarrhea [102]. Edema, neutrophil infiltrate, and reduced enzymatic activity in the intestinal mucosa have been also found in EPEC infection [103]. EPEC diarrhea often lasts 1–2 weeks but can become persistent, lasting more than 2 weeks, and may result in severe infection [24, 25, 32, 102]. In a recent case-control study, EPEC infection was associated with a 2.8-fold elevated risk of death among infants in Kenya [24, 31, 97].

9.2 Treatment

Treatment of EPEC diarrhea includes oral rehydration therapy to prevent dehydration by correcting fluid and electrolyte losses. Oral rehydration may be sufficient for cases of self-limited acute diarrhea, but persistent cases of diarrhea may include parenteral rehydration, and more severe cases may require total parental nutrition and use of antimicrobials [102]. Multidrug resistance has been reported in EPEC strains from diverse parts of the world [27–29, 44, 104, 105]. Alternative therapies, employing the use of bismuth subsalicylate, specific bovine anti-EPEC milk immunoglobulins, and also zinc, have been proven useful for treatment and prevention of EPEC diarrhea [106].

9.3 Vaccines

There are no currently available vaccines to prevent EPEC infection. However, a recent study has used bacterial ghosts devoid of cytoplasmic contents but expressing all EPEC surface components in vaccination challenge experiments with mice, and the results showed 84–90% protection in control mice [107]. Interestingly, protective effect of breast-feeding was shown to provide excellent protection against EPEC infection. Several investigators have shown that breast milk provides protection against EPEC O antigens and outer membrane proteins [108, 109]. Furthermore, IgA antibodies against BFP, intimin, EspA, and EspB proteins were identified in maternal colostrum and serum samples [110–118].

10. Conclusion

Much progress has been made in the last 20 years toward understanding the pathogenesis of EPEC. It has been shown that typical EPEC are still important pathogens associated with severe outbreaks of infant diarrhea, and atypical EPEC are emerging pathogens associated with sporadic outbreaks at all ages worldwide.

Author details

Isabel Cristina Affonso Scaletsky
Departamento de Microbiologia, Imunologia e Parasitologia, Universidade Federal de São Paulo, Brazil

*Address all correspondence to: scaletsky@unifesp.br

References

[1] Lesage AA. Contribution á l´étude des enteritis infantiles-serodiagnostic des races de *Bacterium coli*. Comptes Rendues des Societe Biologique (Paris). 1897;**49**:900-901

[2] Bray J. Isolation of antigenically homogeneous strains of *Bacterium coli neopolitanum* from summer diarrhoea of infants. The Journal of Pathology and Bacteriology. 1945;**57**:239-247

[3] Kauffmann F, Dupon A. *Escherichia coli* strains from infantile epidemic gastroenteritis diarrhea. Acta Neuropathologica. 1950;**27**:221-226

[4] Neter E, Westphal O, Luderitz O, et al. Demonstration of antibodies against enteropathogenic *Escherichia coli* in sera of children of various ages. Pediatrics. 1955;**16**:801-807

[5] Levine MM, Bergquist EJ, Nalin DR, et al. *Escherichia coli* strains that cause diarrhea but do not produce heat-labile or heat-stable enterotoxins and are non-invasive. Lancet. 1978;**1**(8074):1119-1122

[6] Edelman R, Levine MM. From the National institute of allergy and infectious diseases. Summary of a workshop on enteropathogenic *Escherichia coli*. The Journal of Infectious Diseases. 1983;**147**:1108-1118

[7] Cravioto A, Gross RJ, Scotland SM, Rowe B. An adhesive factor found in strains of *Escherichia coli* belonging to traditional infantile enteropathogenic serotypes. Current Microbiology. 1979;**3**:95-99

[8] Scaletsky IC, Silva ML, Trabulsi LR. Distinctive patterns of adherence of enteropathogenic *Escherichia coli* to HeLa cells. Infection and Immunity. 1984;**45**:534-536

[9] Baldini MM, Kaper JB, Levine MM, et al. Plasmid mediated adhesion of enteropathogenic *Escherichia coli*. Journal of Pediatric Gastroenterology and Nutrition. 1983;**2**:534-538

[10] Nataro JP, Baldini MM, Kaper JB, Black RE, Bravo N, Levine MM. Detection of an adherence factor of enteropathogenic *Escherichia coli* with a DNA probe. The Journal of Infectious Diseases. 1985a;**152**:560-565

[11] Nataro JP, Scaletsky ICA, Kaper JB, Levine MM, Trabulsi LR. Plasmid-mediated factors conferring diffuse and localized adherence of enteropathogenic *Escherichia coli*. Infection and Immunity. 1985b;**48**:378-383

[12] Nataro JP, Maher KO, Mackie P, Kaper JB. Characterization of plasmids encoding the adherence factor of enteropathogenic *Escherichia coli*. Infection and Immunity. 1987;**55**:2370-2377

[13] Moon HW, Whipp SC, Argenzio RA, et al. Attaching and effacing activities of rabbit and human enteropathogenic *Escherichia coli* in pig and rabbit intestines. Infection and Immunity. 1983;**41**:1340-1351

[14] Knutton S, Baldini, MM, Kaper JB, McNeish AS. Role of plasmid-encoded adherence factors in adhesion of enteropathogenic *Escherichia coli* to HEp-2 cells. Infection and Immunity. 1987;**55**:78-85

[15] Kaper JB. Defining EPEC. Revista de Microbiologia, Sao Paulo. 1996;**27**:130-133

[16] Ørskov F, Whittam TS, Cravioto A, Ørskov I. Clonal relationships among classic enteropathogenic *Escherichia coli* (EPEC) belong to different O groups. The Journal of Infectious Diseases. 1990;**162**:76-81

[17] Whittam TS, McGraw EA. Clonal analysis of EPEC serogroups. Revista de Microbiologia. 1996;**27**:7-16

[18] Dulguer MV, Fabricotti SH, Bando SY, et al. Atypical enteropathogenic *Escherichia coli* strains: Phenotypic and genetic profiling reveals a strong association between enteroaggregative *E. coli* heat-stable enterotoxin and diarrhea. The Journal of Infectious Diseases. 2003;**188**:1685-1694

[19] Trabulsi LR, Keller R, Gomes TAT. Typical and atypical enteropathogenic *Escherichia coli*. Emerging Infectious Diseases. 2002;**8**:508-513

[20] Vieira MA, Andrade JR, Trabulsi LR. Phenotypic and genotypic characteristics of *Escherichia coli* strains of non-enteropathogenic *E. coli* (EPEC) serogroups that carry EAE and lack the EPEC adherence factor and Shiga toxin DNA probe sequences. The Journal of Infectious Diseases. 2001;**183**:762-772

[21] Hazen TH, Sahl JW, Fraser CM, et al. Refining the pathovar paradigm via phylogenomics of the attaching and effacing *Escherichia coli*. PNAS. 2013;**110**:12810

[22] Ochoa TJ, Barletta F, Contreras C, et al. New insights into the epidemiology of enteropathogenic *Escherichia coli* infection. Transactions of the Royal Society of Tropical Medicine and Hygiene. 2008;**102**:852-856

[23] Maranhão HS, Medeiros MCC, Scaletsky ICA, et al. The epidemiological and clinical characteristics and nutritional development of infants with acute diarrhea, in northeastern Brazil. Annals of Tropical Medicine and Parasitology. 2008;**102**:357-365

[24] Nataro JP, Kaper JB. Diarrheagenic *Escherichia coli*. Clinical Microbiology Reviews. 1998;**11**:142-201

[25] Hu J, Torres AG. Enteropathogenic *Escherichia coli*: Foe or innocent bystander? Clinical Microbiology and Infection. 2015:1-6

[26] Franzolin MR, Alves RCB, Keller R, Gomes TAT, Beutin L, Barreto ML, et al. Prevalence of diarrheagenic *Escherichia coli* in children with diarrhea in Salvador, Bahia, Brazil. Memórias do Instituto Oswaldo Cruz. 2005;**100**:359-363

[27] Scaletsky ICA, Souza TB, Aranda KRS, Okeke IN. Genetic elements associated with antimicrobial resistance in enteropathogenic *Escherichia coli* (EPEC) from Brazil. BMC Microbiology. 2010;**10**:25

[28] Scaletsky ICA, Aranda KR, Souza TB, Silva NP, Morais MB. Evidence of pathogenic subgroups among atypical enteropathogenic *Escherichia coli* strains. Journal of Clinical Microbiology. 2010;**47**:3756-3759

[29] Scaletsky ICA, Aranda KRS, Souza TB, Silva NP. Adherence factors in atypical enteropathogenic *Escherichia coli* strains expressing the localized adherence-like pattern in HEp-2 cells. Journal of Clinical Microbiology. 2010;**48**:302-306

[30] Alikhani MY, Mirsalehian A, Aslani MM. Detection of typical and atypical enteropathogenic *Escherichia coli* (EPEC) in Iranian children with and without diarrhea. Journal of Medical Microbiology. 2006;**55**:1159-1163

[31] Kotloff KL, Nataro JP, Blackwelder W, et al. Burden and aetiology of diarrhoeal disease in infants and young children in developing countries (The Global Enteric Multicentre Study, GEMS): A prospective, case-control study. Lancet. 2013;**382**(9888):209-222

[32] Levine MM, Edelman R. Enteropathogenic *Escherichia coli* of classic serotypes associated with infant diarrhea: epidemiology and

pathogenesis. Epidemiology Review. 1984;**6**:31-51

[33] Behiry IK, Abada EA, Ahmed EA, Labeed RS. Enteropathogenic *Escherichia coli* associated with diarrhea in children in Cairo, Egypt. Scientific World Journal. 2011;**11**:2613-2619

[34] Hernandes RT, Elias WP, Vieira AM, Gomes TAT. An overview of atypical enteropathogenic *Escherichia coli*. FEMS Microbiology Letters. 2009;**297**:137-149

[35] Kolenda R, Burdukiewicz M, Schierck P. A systematic review and meta-analysis of the epidemiology of pathogenic *Escherichia coli* of calves and the role of calves as reservoirs for human pathogenic *E. coli*. Frontiers in Cellular and Infection Microbiology. 2015;**5**:23

[36] Fagundes-Neto U, Freymuller E, Gatti MSV, et al. Enteropathogenic *Escherichia coli* O111ab:H2 penetrates the small bowel epithelium in an infant with acute diarrhea. Acta Paediatrica. 1995;**84**:453-455

[37] Rothbaum R, McAdams AJ, Giannella R, Partin JC. A clinicopathological study of enterocyte-adherent *Escherichia coli*: A cause of protracted diarrhea in infants. Gastroenterologia. 1982;**83**:441-454

[38] Girón JA, Ho AS, Schoolnik GK. An inducible bundle-forming pilus of enteropathogenic *Escherichia coli*. Science. 1991;**254**:710-713

[39] ICA S, Pelayo JS, Giraldi R, et al. EPEC adherence to HEp-2 cells. Proceedings of the International Symposium on Enteropathogenic *Escherichia coli* (EPEC). Revista de Microbiologia. 1996;**27**:58-62

[40] Scaletsky ICA, Pedroso MZ, Oliva CAG, et al. A localized adherence-like pattern as a second pattern of adherence of classic enteropathogenic *Escherichia coli* to HEp-2 cells that is associated with infantile diarrhea. Infection and Immunity. 1999;**67**:3410-3415

[41] Staley TE, Jones EW, Corley LD. Attachment and penetration of *Escherichia coli* into intestinal epithelium of the ileum in newborn pigs. The American Journal of Pathology. 1969;**56**:371-392

[42] Fletcher JN, Embaye HE, Getty B, et al. Novel invasion determinant of enteropathogenic *Escherichia coli* plasmid pLV501 encodes the ability to invade intestinal epithelial cells and HEp-2 cells. Infection and Immunity. 1992;**60**:2229-2236

[43] Scaletsky ICA, Gatti MSV, Da Silveira, et al. Plasmid encoding for drug resistance and invasion of epithelial cells in enteropathogenic *Escherichia coli* O111:H. Microbial Pathogenesis. 1995;**18**:387-399

[44] Croxen MA, Law RJ, Scholz R, et al. Recent advances in understanding enteric pathogenic *Escherichia coli*. Clinical Microbiology Reviews. 2013;**26**:822-880

[45] Moreira CG, Palmer K, Whiteley M, et al. Bundle-forming pili and EspA are involved in biofilm formation by enteropathogenic *Escherichia coli*. Journal of Bacteriology. 2006;**188**:3952-3961

[46] Culler HF, Mota CM, Abe CM, et al. Atypical enteropathogenic *Escherichia coli* strains form biofilm on abiotic surfaces regardless of their adherence pattern on cultured epithelail cells. BioMed Research International. 2014;**26**:822-880

[47] Nascimento HH, Silva LEP, Souza RT, et al. Phenotypic and genotypic characteristics associated with biofilm formation in clinical isolates of atypical enteropathogenic *Escherichia coli* (aEPEC) strains. BMC Microbiology. 2014;**14**:184

[48] Hernandes RT, Dela Cruz MA, Yamamoto D, et al. Dissection of the role of pili and type 2 and 3 secretion systems in adherence and biofilm formation of an atypical enteropathogenic *Escherichia coli* strain. Infection and Immunity. 2013;**81**:3793-3802

[49] Weiss-Muzkat M, Shakh D, Zhou Y, et al. Biofilm formation and multicellular behavior in *Escherichia coli* O55:H7, an atypical enteropathogenic strain. Applied and Environmental Microbiology. 2010;**76**:1545-1554

[50] Brinkley C, Burland V, Keller R, et al. Nucleotide sequence analysis of the enteropathogenic *Escherichia coli* adherence factor plasmid pMAR7. Infection and Immunity. 2006;**74**:5408-5413

[51] Tobe T, Hayashi T, Han C, et al. Complete DNA sequence and structural analysis of the enteropathogenic *Escherichia coli* adherence factor plasmid. Infection and Immunity. 1999;**67**:5455-5462

[52] Bieber D, Ramer SW, Wu CY, et al. Type IV pili, transient bacterial aggregates, and virulence of enteropathogenic *Escherichia coli*. Science. 1998;**280**:2114-2118

[53] Hazen TH, Kaper JB, Nataro JP, Rasko DA. Comparative genomics provides insight into the diversity of the attaching and effacing *Escherichia coli* virulence plasmids. Infection and Immunity. 2015;**83**:4103-4117

[54] Stone KD, Zhang H, Carlson LK, Donnenberg MS. A cluster of fourteen genes from enteropathogenic *Escherichia coli* is sufficient for the biogenesis of a type IV pilus. Molecular Microbiology. 1996;**20**:325-337

[55] Bortoloni M, Trabulsi L, Keller R, et al. Lack of expression of bundle-forming pili in some clinical isolates of enteropathogenic *Escherichia coli* (EPEC) is due to a conserved large deletion in the *bfp* operon. FEMS Microbiology Letters. 1999;**179**:169-174

[56] Teixeira NB, Rojas TCG, Silveira WD, et al. Genetic analysis of enteropathogenic *Escherichia coli* (EPEC) adherence factor (EAF) plasmid reveals a new deletion within the EAF probe sequence among O119 typical EPEC strains. BMC Microbiology. 2015;**15**:200

[57] McDaniel TK, Jarvis KG, Donnenberg MS, Kaper JB. A genetic locus of enterocyte effacement conserved among diverse enterobacterial pathogens. Proceedings of the National Academy of Sciences of the United States of America. 1995;**92**:1664-1668

[58] Dean P, Kenny B. The effector repertoire of enteropathogenic *E. coli* ganging up on the host cell. Current Opinion in Microbiology. 2009;**12**:101-109

[59] Deng W, Puente JL, Grunheid S, et al. Dissecting virulence: Systematic and functional analyses of a pathogenicity island. Proceedings of the National Academy of Sciences of the United States of America. 2004;**101**:3597-3602

[60] Elliott SJ, Sperandio V, Giron JA, et al. The locus of enterocyte effacement (LEE)-encoded regulator controls expression of both LEE- and non-LEE-encoded virulence factors in enteropathogenic *Escherichia coli*. Infection and Immunity. 2000;**68**:6115-6126

[61] Barba J, Bustamante VH, Flores-Valdez MA, et al. A positive regulatory loop controls expression of the locus of enterocyte effacement-encoded regulators Ler and GrlA. Journal of Bacteriology. 2005;**187**:7918-7730

[62] Kenny B, DeVinney R, Stein M, et al. Enteropathogenic *Escherichia coli* (EPEC) transfers its receptor for intimate adherence into mammalian cells. Cell. 1997;**91**:511-520

[63] Santos A, Finlay BB. Bringing down the host: Enteropathogenic and enterohaemorrhagic *Escherichia coli* effector-mediated subversion of host innate immune pathways. Cellular Microbiology. 2015;**17**:318-332

[64] Wong ARC, Pearson JS, Bright MD, et al. Enteropathogenic and enterohaemorrhagic *Escherichia coli*: Even more subversive elements. Molecular Microbiology. 2011;**80**:1420-1438

[65] Vossenkämper A, MacDonald TT, Marches O, et al. Always one step ahead: How pathogenic bacteria use the type III secretion system to manipulate the intestinal mucosal immune system. Journal of Inflammation. 2011;**8**:11

[66] Gärtner JF, Schmidt MA, Ga JF, et al. Comparative analysis of locus of enterocyte effacement pathogenicity islands of atypical enteropathogenic *E. coli*. Society. 2004;**72**:6722-6728

[67] Ingle DJ, Tauschek M, Edwards DJ, et al. Evolution of atypical enteropathogenic *E. coli* by repeated acquisition of LEE pathogenicity island variants. Nature Microbiology. 2016;**1**:15010

[68] Müller D, Benz I, Liebchen A, et al. Comparative analysis of the locus of enterocyte effacement and its flanking regions. Infection and Immunity. 2009;**77**:3501-3513

[69] Sperandio V, Li CC, Kaper JB. Quorum-sensing *Escherichia coli* regulator A: A regulator of the LysR family involved in the regulation of the locus of enterocyte effacement pathogenicity island in enterohemorrhagic *E. coli*. Infection and Immunity. 2002;**70**:3085-3093

[70] Frankel G, Candy DCA, Everest P, Dougan G. Characterization of the C-terminal domains of intimin-like proteins of enteropathogenic and enterohemorrhagic *Escherichia coli*, *Citrobacter freundii*, and *Hafnia alvei*. Infection and Immunity. 1994;**62**:1835-1842

[71] Lacher DW, Steinsland H, Whittam TS. Allellic subtyping of the intimin locus (*eae*) of pathogenic *Escherichia coli* by fluorescent RFLP. FEMS Microbiology Letters. 2006;**261**:80-87

[72] Lacher DW, Steinsland H, Blank TE, et al. Molecular evolution of typical enteropathogenic *Escherichia coli*: Clonal analysis by multilocus sequence typing and virulence gene allelic profiling. Journal of Bacteriology. 2007;**189**:342-350

[73] Ruchaud-Sparagano M-H, Muhlen S, Dean P, Kenny B. The enteropathogenic *Escherichia coli* (EPEC) Tir effector inhibits NF-κB activity by targeting TNFα-receptor-associated factors. PLoS Pathogens. 2011;**7**:e1002414

[74] Girón JA, Ho ASY, Schoolnik GK. Characterization of fimbriae produced by enteropathogenic *Escherichia coli*. Journal of Bacteriology. 1993a;**175**:7391-7403

[75] Girón JA, Donnenberg MS, Martin WC, Jarvis KG, Kaper JB. Distribution of the bundle-forming pilus structural genes (*bfpA*) among enteropathogenic *Escherichia coli*. The Journal of Infectious Diseases. 1993b;**168**:1037-1041

[76] Badea L, Doughty S, Nicholls L, et al. Contribution of Efa/LifA to the adherence of enteropathogenic *Escherichia coli* to epithelial cells. Microbial Pathogenesis. 2003;**34**:205-215

[77] Klapproth JM, Scaletsky ICA, McNamara BP, et al. A large toxin from pathogenic *Escherichia coli* strains that inhibits lymphocyte activation. Infection and Immunity. 2000;**68**:2148-2155

[78] Klapproth JM, Sasaki M, Sherman M, et al. *Citrobacter rodentium lifA/efa1* is essential for colonic colonization and crypt cell hyperplasia *in vivo*. Infection and Immunity. 2005;**73**:1441-1451

[79] Vieira MAM, Salvador FA, Silva RM, et al. Prevalence and characteristics of the O122 pathogenicity island in typical and atypical enteropathogenic *Escherichia coli* strains. Journal of Clinical Microbiology. 2010;**48**:1452-1455

[80] Afset JE, Bruant G, Brousseau R, et al. Identification of virulence genes linked with diarrhea due to atypical enteropathogenic *Escherichia coli* by DNA microarray analysis and PCR. Journal of Clinical Microbiology. 2006;**44**:3703-3711

[81] Torres AG, Tutt CB, Durval L, et al. Bile salts induce expression of the afimbrial LDA adhesion of atypical enteropathogenic *Escherichia coli*. Cellular Microbiology. 2007;**9**:1039-1049

[82] Saldaña Z, Erdem AL, Schuller S, et al. The *Escherichia coli* common pilus and the bundle-forming pilus act in concern during the formation of localized adherence by enteropathogenic *E. coli*. Journal of Bacteriology. 2009;**191**:3451-3461

[83] Girón JA, Torres AG, Freer E, et al. The flagella of enteropathogenic *Escherichia coli* mediate adherence to epithelial cells. Molecular Microbiology. 2002;**44**:361-479

[84] Clearly J, Lai LC, Shaw RK, et al. Enteropathogenic *Escherichia coli* (EPEC) adhesion to intestinal epithelial cells: Role of bundle-forming pili (BFP), EspA filaments and intimin. Microbiology. 2004;**150**:527-538

[85] Vidal JE, Navarro-Garcia F. EspC translocation into epithelial cells by enteropathogenic *Escherichia coli* requires a concerted participation of type V and III systems. Cellular Microbiology. 2008;**10**:1976-1986

[86] Drago-Serrano ME, Parra SG, Manjarrez-Hernández HA. EspC, an autotransporter protein secreted by enteropathogenic *Escherichia coli* (EPEC), displays protease activity on human hemoglobin. FEMS Microbiology Letters. 2006;**265**:35-40

[87] Salinger N, Kokona B, Fairman R, Okeke I. The plasmid encoded regulator activates factors conferring lysozyme resistance on enteropathogenic *Escherichia coli* strains. Applied and Environmental Microbiology. 2009;**75**:275-280

[88] Xicohtencatl-Cortes J, Saldaña Z, Deng W, et al. Bacterial macroscopic ropelike fibers with cytopathic and adhesive properties. The Journal of Biological Chemistry. 2010;**285**:32336-32342

[89] Scott DA, KaperJB. Cloning and sequencing of the genes encoding *Escherichia coli* cytolethal distending toxin. Infection and Immunity. 1994;**62**:244-251

[90] Lara-Tejero M, Galan JE. A bacterial toxin that controls cell cycle progression as a deoxyribonuclease I-like protein. Science. 2000;**290**:354-357

[91] Albert MJ, Faruque SM, Faruque ASG, et al. Controlled study of cytolethal distending toxin-producing Escherichia coli infections in Bangladeshi children. Journal of Clinical Microbiology. 1996;**34**:717-719

[92] Yamamoto T, Wakisaka N, Sato F, Kato A. Comparison of the nucleotide

sequence of enteroaggregative *Escherichia coli* heat-stable enterotoxin 1 genes among diarrhea-associated *Escherichia coli*. FEMS Microbiology Letters. 1997;**147**:89-96

[93] Silva LEP, Souza TB, Silva NP, Scaletsky ICA. Detection and genetic analysis of the enteroaggregative *Escherichia coli* heat-stable enterotoxin (EAST1) gene in clinical isolates of enteropathogenic *Escherichia coli* (EPEC) strains. BMC Microbiology. 2014;**14**:135

[94] Donnenberg MS, Kaper JB. Enteropathogenic *Escherichia coli*. Infection and Immunity. 1992;**60**:3953-3961

[95] Clarke SC et al. Virulence of enteropathogenic *Escherichia coli*, a global pathogen. Clinical Microbiology Reviews. 2003;**6**:365-378

[96] Campellone KG, Giese A, Tipper DJ, Leong JM. A tyrosine-phosphorylated 12-aminoacid sequence of enteropathogenic *Escherichia coli* Tir binds the host adaptor protein Nck and is required for Nck localization to actin pedestals. Molecular Microbiology. 2002;**43**:1227-1241

[97] Goosney DL, DeVinney R, Finlay BB. Recruitment of cytoskeletal and signaling proteins to enteropathogenic and enterohemorrhagic *Escherichia coli* pedestals. Infection and Immunity. 2001;**69**:3315-3322

[98] Knutton S, Baldwin T, Williams PH, McNeish AS. Actin accumulation at sites of bacterial adhesion to tissue culture cells: Basis of a new diagnostic test for enteropathogenic and enterohemorrhagic *Escherichia coli*. Infection and Immunity. 1989;**57**:1290-1298

[99] Franke J, Franke S, Schmidt H, et al. Nucleotide sequence analysis of enteropathogenic *Escherichia coli* (EPEC) adherence factor probe and development of PCR for rapid detection of EPEC harboring virulence plasmids. Journal of Clinical Microbiology. 1994;**32**:2460-2463

[100] Gunzburg ST, Tornieporth NG, Riley LW. Identification of enteropathogenic *Escherichia coli* by PCR-based detection of the bundle-forming pilus gene. Journal of Clinical Microbiology. 1995;**33**:1375-1377

[101] Blank TE, Zhong H, Bell AL, et al. Molecular variation among type IV pilin (*bfpA*) genes from diverse enteropathogenic *Escherichia coli* strains. Infection and Immunity. 2000;**68**:7028-7038

[102] Fagundes-Neto U, Scaletsky ICA. The gut at war: The consequences of enteropathogenic *Escherichia coli* infection as a factor of diarrhea and malnutrition. São Paulo Medical Journal/Revista Paulista de Medicina. 2000;**118**:21-29

[103] Arenas-Hernandez MM, Martinez-Laguna Y, Torres AG. Clinical implications of enteroadherent *Escherichia coli*. Current Gastroenterology Reports. 2012;**14**:386-394

[104] Mitra M, Ahmad P, Mehdi R, Hosein A, Ahmad K. Multiple drug resistance of enteropathogenic *Escherichia coli* isolated from children with diarrhea in Kashan, Iran. African Journal of Microbiology Research. 2011;**5**:3305-3309

[105] Subramanian K, Selvakkumar C, Vinaykumar KS, et al. Tackling multiple antibiotic resistance in enteropathogenic *Escherichia coli* (EPEC) clinical isolates: A diarylheptanoid from Alpinia officinarum shows promising antibacterial and imunomodulatory activity against EPEC and its lipopolysacharide-induced inflammation. International Journal of Antimicrobial Agents. 2009;**33**:244-250

[106] Patel A, Mamtani M, Dibley MJ, et al. Therapeutic value of zinc supplementation in acute and persistent diarrhea: A systematic review. PLoS One. 2010;**5**:e10386

[107] Liu J, Wang WD, Liu YJ, et al. Mice vaccinated with enteropathogenic *Escherichia coli* ghosts show significant protection against lethal challenges. Letters in Applied Microbiology. 2012;**54**:255-262

[108] Costa-Carvalho BT, Bertipaglia A, Solé D, et al. Detection of immunoglobulin (IgG and IgA) anti-outer membrane proteins of enteropathogenic *Escherichia coli* (EPEC) in saliva, colostrum, breast milk, serum, cord blood and amniotic fluid. Study of inhibition of localized adherence of EPEC to HeLa cells. Acta Paediatrica. 1994;**83**:870-873

[109] Cravioto A, Tello A, Villafan H, et al. Inhibition of localized adhesion of enteropathogenic *Escherichia coli* to HEp-2 cells by immunoglobulin and oligosaccharide fractions of humancolostrum and breast milk. The Journal of Infectious Diseases. 1991;**163**:1247-1255

[110] Fernandes RCSC, Quintana Flores VM, Sousa de Macedo Z, Medina-Acosta E. Coproantibodies to the enteropathogenic *Escherichia coli* vaccine candidates BfpA and EspB in breast fed and artificially fed children. Vaccine. 2003;**21**:1725-1731

[111] Parissi-Crivelli A, Parissi-Crivelli JM, Girón JA. Recognition of enteropathogenic *Escherichia coli* virulence determinants by human colostrum and serum antibodies. Journal of Clinical Microbiology. 2000;**38**:2696-2700

[112] Quintana Flores VM, Fernandes RCCS, Sousa de Macedo Z, Medina-Acosta E. Expression and purification of the recombinant enteropathogenic *Escherichia coli* vaccine candidates BfpA and EspB. Protein Expression and Purification. 2002;**25**:16-22

[113] Sanches MI, Keller R, Hartland EL, et al. Human colostrum and serum contain antibodies reactive to the intimin-binding region of the enteropathogenic *Escherichia coli* translocated intimin receptor. Journal of Pediatric Gastroenterology and Nutrition. 2000;**30**:73-77

[114] Stakenborg T, Vandekerchove D, Mariën J, et al. Protection of rabbits against enteropathogenic *Escherichia coli* (EPEC) using an intimin null mutant. BMC Veterinary Research. 2006;**2**:22

[115] Tacket CO, Sztein MB, Losonsky G, et al. Role of EspB in experimental human enteropathogenic *Escherichia coli* infection. Infection and Immunity. 2000;**68**:3689-3695

[116] Carbonare SB, Silva ML, Palmeira P, Carneiro-Sampaio MM. Human colostrum IgA antibodies reacting to enteropathogenic *Escherichia coli* antigens and their persistence in the faeces of a breastfed infant. Journal of Diarrhoeal Diseases Research. 1997;**15**:53-58

[117] Donnenberg MS, Tacket CO, Losonsky G, et al. Effect of prior experimental human enteropathogenic *Escherichia coli* infection on illness following homologous and heterologous re-challenge. Infection and Immunity. 1998;**66**:52-58

[118] Loureiro I, Frankel G, Adu-Bobie J, et al. Human colostrum contains IgA antibodies reactive to enteropathogenic *Escherichia coli* virulence-associated proteins: Intimin, BfpA, EspA, and EspB. Journal of Pediatric Gastroenterology and Nutrition. 1998;**27**:166-171

Chapter 4

Insight into the mobilome of *Escherichia coli*

Elif Bozcal

Abstract

Mobilomes are all mobile genetic elements (plasmids, transposable elements, insertion sequences, gene cassettes, integrons, genomic islands, and bacteriophages) in a genome. Mobilome is one of the responsible agents for the bacterial evolution, virulence, and increasing antibiotic resistance. The mobile genetic elements in the *Escherichia coli* genome can carry antibiotic resistance genes and/or virulence genes. The acquisition of new mobile genetic elements can lead to the emergence of new pathotypes. The aim of this chapter is to gather knowledge about mobile genetic elements in *E. coli* strains. The method in this chapter depends on a literature survey, which scans reviews, research articles, and theses published about transposable elements, plasmids, bacteriophages, and genomic islands in *E. coli* strains.

Keywords: *Escherichia coli*, mobilome, genomic island, plasmid, transposon, bacteriophage

1. Introduction to mobilome and mobile genetic elements

Mobilome encompasses all mobile genetic elements (MGEs) in a genome [1]. Mobile genetic elements are moveable DNA segments, transferring among bacterial genomes. MGEs carry the so-called noncore genes and they have an important contribution to the plasticity of bacterial genomes. Plasmids, transposable elements, insertion sequences, gene cassettes, integrons, genomic islands, and bacteriophages are MGEs. Approximately, 2000 genes from 20 sequenced *E. coli* genomes were found to be noncore genes [2].

The presence and/or absence of MGEs within genomic DNA can lead to variance in *E. coli* pathotypes. Despite the fact that *E. coli* strains have been known as part of the normal microbiota of human gastrointestinal tract among *E. coli,* there are also pathogenic strains, and hence, the strains of *E. coli* are grouped either as (i) commensal, which are nonpathogenic, (ii) intestinal pathogenic strains (IPEC), or (iii) extraintestinal pathogenic (ExPEC) strains [3]. Rearrangements, excision, and integration of the DNA fragments can be the mechanisms behind the rapid evolution of *E. coli* strains and also the emerging pathogenic *E. coli* strains [4].

The most studied MGEs are transposable elements, plasmids, bacteriophages, and genomic islands in *E. coli* strains. Transposable elements are known as DNA sequences that can transfer among different locations in the bacterial genome [5]. Resistance genes can be harbored by a transposon in a genome. Transposons can be integrated and excised from the chromosome by enzymes, called transposases. The simplest version of the transposon is the insertion sequence [6]. Plasmids are self-replicating genetic elements. Different groups of plasmids exist according to the incompatibility and conjugative features. Plasmids have a big contribution to the bacterial cell in terms of acquiring antibiotic resistance genes and virulence genes [7, 8]. Bacteriophages are viruses that infect bacteria and replicate within bacterial cells. Bacteriophages can transfer genes among bacterial cells with the mechanism, called transduction. Specialized transduction can include only specific genes, however, generalized transduction can transfer any fragment of the bacterial DNA [6]. In a similar manner, some bacteriophages also carry genes, which are advantageous for bacteria such as resistance and virulence-associated genes. Among them, Shiga-toxin coding genes are one of the most significant phage-associated genes that is transferred to *E. coli* O157 : H7 [9]. Important mobile genetic elements are also genomic islands (GIs), which are genomic regions of gene clusters, often acquired by horizontal gene transfer and inserted into tRNA genes [10]. GIs can contain phage- or plasmid-derived sequences. GIs in *E. coli* strains carry genes associated with metabolism, pathogenesis as well as antimicrobial resistance [11].

2. Transposable elements or transposons

Transposable elements (TEs) or transposons have a significant role in the genome evolution and organization [12]. TEs are DNA fragments, which are able to change their position within the genome, in the process called transposition [13]. Bacterial transposable elements or bacterial transposons are divided into three different types: (i) insertion sequence elements, (ii) composite transposons, and (iii) noncomposite transposons. Insertion sequence elements, or in short, insertion sequences (ISs) are the simplest version of the transposable elements. ISs have not genetic information apart from necessary for their mobility. Composite and noncomposite transposons, on the other hand, also they have additional genetic material unrelated with transposition, for example, antibiotic resistance genes [14]. Composite transposons are flanked by the insertion sequences [12]. *E. coli* has various transposable elements carrying antibiotic resistance genes, including Tn*3*, Tn*5*, Tn*7*, Tn*9*, and Tn*10* encoding ampicillin, kanamycin, trimethoprim, chloramphenicol, and tetracycline, respectively [13, 15]. A transposon, Tn*6306*, encoding imipenem-hydrolyzing β-lactamase that mediates dissemination of the $blaIM_{I}$ among *Enterobacteriaceae* reported in 2017 [16]. The gene conferring resistance

Name	Description	Reference
Transposon-like element	bla_{CMY-2}-carrying element	[22]
Tn*21*-type transposon	*dfrA* trimethoprim resistance	[18]
Tn*1999*-like element	bla_{OXA-48} gene encoding OXA-48 carbapenem	[23]
Tn*6306*	bla_{IMI}	[16]
ISApl1	*mcr-1*	[17]
In*53*	bla_{VEB-1}	[21]
Tn*3*	Ampicillin resistance	[24]
Tn*5*	Kanamycin resistance	[25]
Tn*7*	Trimethoprim resistance	[26]
Tn*9*	Chloramphenicol resistance	[27]
Tn*10*	Tetracycline resistance	[28]

Table 1.
Transposable elements in E. coli *strains.*

to colistin, *mcr-1* gene is carried by the ISApl1 transposon [17]. Moreover, trimethoprim resistance gene *dfrA* was rapidly disseminated in Nigeria and Ghana via Tn*21*-type transposon in *E. coli* [18]. In **Table 1**, the most important transposable elements of *E. coli* are listed.

Sometimes transposable elements comprise integrons, which are genetic elements that can capture genes including antibiotic resistance from different sources [19]. And integrons can be located on transposons, but also on plasmids, and in the bacterial chromosome [20]. Integrons are genetic elements that include the gene for the enzyme integrase together with gene cassettes encoding antibiotic resistance genes. A study reported a novel integron, In*53*, which is located on transposon inserted into a self-transferable plasmid [21].

3. Plasmids

Plasmids are extrachromosomal DNA elements, which are self-replicating. Apart from the genetic information needed for the autonomous replication, they can also carry additional genetic information like antibiotic resistance genes and the genes encoding resistance to heavy metals, virulence, and other metabolic functions [22, 29]. Thanks to their specific functions, certain plasmids are used as cloning vectors in the recombinant DNA technology [30]. Plasmids are grouped into different Inc families/groups. Inc groups are based on the inability of two plasmids to co-exist together in a bacterial cell [31]. Inc plasmids of the same Inc group have same type of replication region and thus have incompatible replication and partition mechanisms and hence cannot co-exist in a bacterial cell [32]. Plasmids belonging to the IncX family encode various resistance genes, mainly distributed among members of *Enterobacteriaceae* [33]. For example, recently, an emerging IncX plasmid, which is encoding $bla_{SHV\text{-}12}$ β-lactamase gene was reported in *E. coli*. The $bla_{IMI\text{-}2}$ gene, encoding an imipenem-hydrolyzing β-lactamase, is carried by pRJ18, an IncFIB plasmid [16]. The ESBL-encoding plasmids belonging to the Inc F, A/C, N, H12, 11 and K type were reported from The European Union. And one of the significant ESBL enzyme genes, CTX-M-1, is generally located on the Inc1 or IncN plasmids. For instance, CTX-M-1 ß-lactamase originated from an animal is transmitted via Inc1 ST3 plasmid [34].

The paradigm F plasmid, found among members of *Enterobacteriaceae,* is an IncF plasmid [35]. F-like plasmids can be detected in pathogenic as well as in non-pathogenic *E. coli* strains of different origins. The whole genome sequencing data of *E. coli* ST131 indicated that the acquisition of the CTX-M resistance gene was transferred via the conjugative F plasmid [36]. However, with the F plasmid, another significant antibiotic resistance gene is transferred, the *mcr-1* gene conferring resistance to colistin. Moreover, this *mcr-1* gene was found to be carried by 13 different plasmid incompatibility groups, among them are the Incl2, IncX4, and IncHI2 [37]. Although it has been reported that *mcr-1* gene has been carried by a transposon, it was shown that the *mcr-1* gene was transported via the plasmid, firstly. Moreover, it was found that the other *mcr* genes including *mcr-2, mcr-3, mcr-4, mcr-5,* and variants are carried by plasmids [38–40]. Recently, a new international outbreak was reported in Denmark. This outbreak was caused by *E. coli* ST410, an extraintestinal pathogenic *E. coli* resistant to fluoroquinolones, third-generation, cephalosporins, and carbapenems. This strain has acquired an IncX3 plasmid carrying $bla_{OXA\text{-}181}$ gene and also an IncFII plasmid carrying $bla_{NDM\text{-}5}$ [41].

Sometimes, plasmids are transferred among bacteria with conjugation, a genetic transfer that occurs between donor and recipient cell that is in a direct cell-to-cell contact [42]. Conjugative plasmids can carry integrons and/or transposons, and such genetic information can be transferred then horizontally via conjugation [43].

Therefore, the spreading of multiresistant genes is promoted [44]. For example, a conjugative *E. coli* plasmid from a swine incorporated a *cfr* gene, which conferred resistance to phenicol, lincosamides, oxazolidinones, pleuromutilins, streptogramin A, and also the $bla_{CTX\text{-}M\text{-}14b}$ ESBL gene [45]. Moreover, a colV plasmid (pCERC3) from a commensal *E. coli* ST95 carried virulence and antibiotic resistance genes including sulfonamide resistance encoded by sul3-associated with a class 1 integron [46]. The pE80 plasmid from a foodborne *E. coli* strain encodes multiple resistance determinants *oqxAB*, *fosA3*, *blaCTX-M-55*, and *blaTEM-1* and hence confers resistance to tetracycline, streptomycin, olaquindox/quinolone, and kanamycin [47].

In addition to carrying antibiotic resistance genes, plasmids have a major role in the transfer of virulence-associated genes. One of the most significant *E. coli* outbreaks was the hybrid enterohemorrhagic *E. coli* (EHEC)-enteroaggregative *E. coli* (EAEC) O104:H4 strain in Germany. This strain possesses three different plasmids: pAA (7.4 kb), pESBL (89 kb), and pG (1.5 kb) [48, 49]. pAA plasmid carries virulence factors including fimbriae for adherence, surface protein dispersion, Aat

Name or class	Description/gene carried	Reference
IncA/C plasmids	Multiple antibiotic resistance	[29]
IncX3 plasmid	$bla_{SHV\text{-}12}$	[33]
IncA/C or IncI1 plasmid	$bla_{CMY\text{-}2}$-like genes	[52]
pRJ18	$bla_{IMI\text{-}2}$	[16]
Incl2, IncX4, and IncHI2 plasmid	*mcr-1*	[37]
MOBP131/IncL plasmids	$bla_{OXA\text{-}48}$	[23]
IncX3 plasmid	$bla_{OXA\text{-}181}$	[41]
IncFII plasmid	$bla_{NDM\text{-}5}$	[41]
pEC26	*mcr-1.9*	[53]
pCERC3	A colV plasmid carrying Sul3-related integron	[46]
pAA	Encoding aggregative adherence fimbriae variant	[48]
	aap gene encoding the surface protein dispersin	[50]
	aatPABCD operon encoding Aat secretion system	[50]
	Protease SepA	[50]
	aggR gene is encoding EAEC master virulence gene regulator AggR	[50]
	Support of the translocation of the Stx2a across the epithelial cell	[54]
pS88	Encoding virulence genes $ompT_p$, *sitA*, *cia*, *iss*, *iroN*, *hlyF*, *cvaA*	[55]
pO157	Encoding enterohemolysin, serine protease, type II secretion protein, catalase, peroxidase, toxinB	[8]
pED1169	Encoding F4-like fimbriae	[7]
pGXEC3	Conjugative plasmid harboring *cfr* gene	[45]
pE80	Conjugative IncFII plasmid encoding *oqxAB*, *fosA3*, $bla_{CTX\text{-}M\text{-}55}$, and *blaTEM-1*	[47]
Incl ST131	$bla_{CTX\text{-}M\text{-}1}$	[34]

Table 2.
Plasmids in E. coli *strains.*

secretion system, protease, and the virulence regulator AggR [50]. Since the pAA plasmid is in the same cell as the pESBL, it increases both virulence and antibiotic resistance of this bacterium. Further, an EHEC O104:H7 strain that was isolated from cattle feces possessed IncB/O/K/Z and IncFIB plasmids carrying principal virulence genes, including, enterohemolysin and autotransporter [51]. Another significant serotype *for E. coli* is the O103 serotype, which is the second most common serogroup among the human foodborne illness. This serogroup has a pO157 plasmid encoding various virulence factors including enterohemolysin and type II secretion protein [8]. Recently, a characterized novel plasmid in a shiga-toxin-producing ETEC harbored *fae* locus encoding ETEC F4 fimbriae [7]. Some more examples of important *E. coli* plasmids are given in **Table 2**.

4. Bacteriophages

Viruses that are infecting bacteria are called bacteriophages. They have a significant impact on the dissemination of antibiotic resistance and virulence-associated genes among foodborne pathogens, as they can transfer genes among bacteria in the process called phage-mediated transduction. Hence, they not only shape the bacterial evolution, but also cause the emergence of new pathogenic bacteria. On the other hand, in a good sense, phages protect against bacterial colonization of mucosal surfaces [56]. Moreover, viruses can be found everywhere in the world including soils, oceans, sewage, and different microbial communities [57, 58]. Transduction can be mediated via virulent or temperate phages. In the case of virulent phages, essentially any region of the bacterial DNA can be transferred (generalized transduction), while temperate phages can transfer only certain genes that are close to the attachment site of the lysogenic phage in the bacterial chromosome (specialized transduction). Specialized transduction happens when the prophage excision is inaccurate and some bacterial DNA co-excised with the prophage DNA [57]. Transduction is a significant process in terms of transferring antimicrobial resistance genes among bacterial cells [59]. For example, the phage called 933E transferred tetracycline resistance gene from the *E. coli* O157:H7 strain to the laboratory *E. coli* K12 AB1157 strain [60]. Similar to plasmids, phages have a crucial role in the acquisition of β-lactamase genes such as $bla_{CTX\text{-}M}$, bla_{SHV}, bla_{TEM}, *qnrA*, *qnrB*, and *qnrS* [61]. A well-characterized P1-like bacteriophage, which lysogenizes bacteria, was reported, and it has encoded SHV-2 resistance in its genomic structure [62].

Furthermore, phages also have the ability to disseminate virulence factors such as staphylokinase, phospholipase or DNase, and superantigens [58]. Phages, which have been known for several years including bacteriophage λ, have been found to carry not only bacterial adhesion genes but also bacterial survival genes [63, 64]. Additionally, *E. coli* phage Ayreon carries the *cdt* gene cluster encoding the CdtA, CdtB, and CdtC subunits of the cdtI holotoxin [65]. Another significant toxin encoded by a temperate phage is the Shiga Toxin 2, which is a virulence factor in *E. coli* O157:H7 [9]. Moreover, some other Shiga Toxin variants including Shiga Toxin 2c can be encoded by phages of the pathogenic *E. coli* O157 strain [66]. However, some bacteriophages, for example, phiC119 can be used as biological control agents, as they can infect and lyse their bacterial hosts (phage therapy) [67]. Interestingly, genetic elements encoded by bacteriophages can also regulate gene expression in the bacterial host cell. For instance, transcription factor Cro has an effect on the regulation of virulence genes in enterohemorrhagic *E. coli* [68]. **Table 3** shows some well-known phages involved in the transduction of virulence or resistance genes.

Name	Description/gene carried	Reference
Bacteriophage 933W	*tet*	[60]
P1-like bacteriophage (RCS47)	bla_{SHV-2}	[62]
Phage ayreon	*cdt* gene cluster	[65]
Phage 2851	*Stx2c*	[66]
Phage λ	*lom* gene encoding K12 adhesion to human buccal epithelial cells	[64]
Phage λ	*bor* gene homologous to *iss* serum resistance locus	[63]
Phage PP01	Two tail fiber genes- 37 and 38 responsible for host-cell recognition	[69]

Table 3.
Bacteriophages of E. coli.

5. Genomic islands

Genomic island (GI) is a large region of genomic DNA, more than 10 kb length, which can be frequently exchanged between bacterial isolates. GIs are encoding proteins for transfer, recombination, and restriction/modification or other proprieties, for example, gene clusters for metabolic adaptation, virulence, and or resistance of bacteria [11]. GIs involving virulence-associated genes are called pathogenicity islands (PAIs) [70]. PAI generally encodes genes related to virulence factors (VFs) including adhesins, toxins, invasins, capsule biosynthesis machinery, iron uptake system, and type III, IV, VI and or VII secretion apparatus [71]. Generally, the size of PAIs more than 10 kb and their GC content differs from the average genome. Their integration site is located in tRNA genes and repeated sequences, which is containing at least one mobile genetic element including remnants of plasmids, insertion sequences (ISs), and integrons, and associated gene cassettes [72]. tRNA-encoding genes are known as the hot spot for the integration of foreign DNA [71]. Several PAIs can be excised from bacterial chromosome by site-specific recombination [73].

PAIs have been reported firstly in the genome of uropathogenic *E. coli*, later also in other pathogenic bacteria [74]. PAIs are now found to be widely distributed among animal- and plant-associated bacterial pathogens and PAIs that can be horizontally transferred have great impact on the rapid evolution of virulent and antibiotic-resistant strains [71].

Enterocyte effacement locus (LEE) is one of the best known PAIs in *E. coli*. LEE is a 35-kb cluster of genes associated with bacterial adherence to intestinal epithelial cells and the formation of attaching and effacing lesions [75]. High-pathogenicity island (HPI), a PAI originally found in *Yersinia* species, but also widely spread among other *Enterobacteriaceae* including *E. coli*, encodes a siderophore iron uptake system (the *fyuA-irp* gene clusters), the so-called yersiniabactin. HPI was found among enteroaggregative *E. coli*, enteropathogenic *E. coli*, enteroinvasive *E. coli*, and enterotoxigenic *E. coli* (ETEC) [76]. HPI was also found in genomes of certain non-O157 STEC clonal lineages [77] (**Table 4**).

It was reported that UPEC strain 536 has at least four PAIs located on the chromosome. The sizes of the first two PAI I536 and PAI II536 are 70 and 120 kb. The significant virulence-associated genes encoded on these PAIs are hemolysin and P fimbriae. PAI III536 and PAI IV536 have S fimbriae and HPI analog gene clusters [78]. It was reported that an ExPEC strain causing neonatal meningitis possessed the HPI, suggesting that HPI was associated with the development of neonatal meningitis [79]. Moreover, capsule synthesis-associated genes can be

Name	Description	Reference
LEE	PAI locus of enterocyte effacement	[75]
Yersiniabactin-HPI	Encoding an siderophore iron uptake system	[77]
PAI I536 and PAI II536	Hemolysin and P fimbriae	[78]
PAI III536 and PAI IV536	S fimbriae and HPI	[78]
PAI V536	K15 capsule determinant	[80]
Vat-encoding PAI	Adjacent to the 3′ terminus-thrW tRNA gene	[82]
EPAI1	RTX family exoprotein	[85]
EPAI2	O-antigen polysaccharide (OPS)	[85]
EPAI3	EPAI3	[85]
EPAI4	T3SS	[85]
EPAI5	O122	[85]
EPAI6	LEE	[85]
AGI-3 PAI	SelC-associated GI involved in carbohydrate uptake and virulence	[86]
GI OI-29	Transcriptional activator GmrA	[83]

Table 4.
Genomic islands and pathogenicity islands in E. coli.

located on different PAIs in ExPEC [80]. Furthermore, it was shown that HPI in general contributes to ExPEC virulence [81]. Another novel PAI was found in an APEC strain integrated adjacent to the thrW tRNA gene encoding vacuolating autotransporter toxin. This PAI is known as Vat-encoding pathogenicity island and may contribute to APEC pathogenicity [82]. A novel function carried by a GI was reported recently—a GI located in the EHEC chromosome included the transcriptional activator GmrA that controls the motility of EHEC O157:H7 [83]. In addition, GIs have a role not only in virulence but also in the metabolic process of the bacterial cell. The *dnd* operon, which is a DNA modification system cluster encoding and catalyzing phosphorothioation of DNA in *E. coli* was found to be located on diverse GIs in *E. coli* [84]. **Table 4** shows some well-known GIs and PAIs.

6. Conclusion

E. coli is one of the most studied bacteria all around the world. There are various pathotypes and subclones of *E. coli* as well as commensal strains. One of the most important reasons of emerging numerous pathotypes of *E. coli* is MGEs, including transposons, plasmids, bacteriophages, and genomic islands. MGEs are significant drivers of the horizontal gene transfer. Therefore, in order to understand the genome evolution, virulence and antibiotic resistance genes acquisition among *E. coli* strains from various sources is important to study MGEs and the mobilome.

Acknowledgements

This study was supported by a fellowship from the Program (TUBITAK-2219-2017-2nd term), and Elif Bozcal was awarded a fellowship by the Scientific and Technological Research Council of Turkey.

Conflict of interest

The author has no conflict of interest.

Author details

Elif Bozcal[1,2]

1 Department of Biology, Basic and Industrial Microbiology Section, Faculty of Science, Istanbul University, Istanbul, Turkey

2 Department of Veterinary and Biomedical Sciences, University of Minnesota, Saint Paul, Minnesota, United States

*Address all correspondence to: elif.bozcal@istanbul.edu.tr

References

[1] Siefert LJ. Defining the Mobilome. In: MBGea, editor. Horizontal Gene Transfer: Genomes in Flux. United States: Humana Press, a part of Springer Science+Business Media. 2009;**532**:13-27

[2] Rankin DJ, Rocha EP, Brown SP. What traits are carried on mobile genetic elements, and why? Heredity. 2011;**106**(1):1-10

[3] Köhler CD, Dobrindt U. What defines extraintestinal pathogenic *Escherichia coli*? International Journal of Medical Microbiology. 2011;**301**(8):642-647

[4] Javadi M, Bouzari S, Oloomi M. Horizontal gene transfer and the diversity of *Escherichia coli*. In: Samie A, editor. *Escherichia coli*—Recent Advances on Physiology, Pathogenesis and Biotechnological Applications. United Kingdom: INTECH; 2017

[5] Munoz-Lopez M, Garcia-Perez JL. DNA transposons: Nature and applications in genomics. Current Genomics. 2010;**11**(2):115-128

[6] Brown-Jaque M, Calero-Caceres W, Muniesa M. Transfer of antibiotic-resistance genes via phage-related mobile elements. Plasmid. 2015;**79**:1-7

[7] Michelacci V, Maugliani A, Tozzoli R, Corteselli G, Chiani P, Minelli F, et al. Characterization of a novel plasmid encoding F4-like fimbriae present in a Shiga-toxin producing enterotoxigenic *Escherichia coli* isolated during the investigation on a case of hemolytic-uremic syndrome. International Journal of Medical Microbiology. 2018;**308**(7):947-955

[8] DebRoy C, Noll LW, Worley JN, Yang X, Shridhar PB, Ludwig JB, et al. Comparative genomics reveals differences in mobile virulence genes of *Escherichia coli* O103 pathotypes of bovine fecal origin. PLoS One. 2018;**13**(2):e0191362

[9] Gamage SD, Strasser JE, Chalk CL, Weiss AA. Nonpathogenic *Escherichia coli* can contribute to the production of Shiga toxin. Infection and Immunity. 2003;**71**(6):3107-3115

[10] Bertelli C, Tilley KE, Brinkman FSL. Microbial genomic island discovery, visualization and analysis. Briefings in Bioinformatics. 2018. [Epub ahead of print]

[11] Hacker J, Carniel E. Ecological fitness, genomic islands and bacterial pathogenicit—A Darwinian view of the evolution of microbes. EMBO Reports. 2001;**2**(5):376-381

[12] Sousa A, Bourgard C, Wahl LM, Gordo I. Rates of transposition in *Escherichia coli*. Biology Letters. 2013;**9**(6):20130838

[13] Iyer A, Barbour E, Azhar E, El Salabi AA, Hassan HM, Qadri I, et al. Transposable elements in *Escherichia coli* antimicrobial resistance. Advances in Bioscience and Biotechnology. 2013;**4**:415-423

[14] Snyder L, Champness W, Champness W. Molecular Genetics of Bacteria. 3rd ed. Washington, DC: ASM Press; 2008

[15] Roberts AP, Chandler M, Courvalin P, Guédon G, Mullany P, Pembroke T, et al. Revised nomenclature for transposable genetic elements. Plasmid. 2008;**60**(3):167-173

[16] Zhang F, Wang X, Xie L, Zheng Q, Guo X, Han L, et al. A novel transposon, Tn 6306, mediates the spread of Bla IMI in *Enterobacteriaceae* in hospitals. International Journal of Infectious Diseases. 2017;**65**:22-26

[17] Wang R, van Dorp L, Shaw LP, Bradley P, Wang Q, Wang X, et al. The global distribution and spread of the mobilized colistin resistance

gene mcr-1. Nature Communications. 2018;**9**(1):1-9

[18] Ibekwe AM, Labar AS, Millman JS, Ruebush E, Opintan JA, Bishar RA, et al. Regional dissemination of a trimethoprim-resistance gene cassette via a successful transposable element. PLoS One. 2012;7(5)

[19] Partridge SR, Brown HJ, Stokes HW, Hall RM. Transposons Tn1696 and Tn21 and their integrons In4 and In2 have independent origins. Antimicrobial Agents and Chemotherapy. 2001;**45**(4):1263-1270

[20] Fluit AC, Schmitz FJ. Resistance integrons and super-integrons. Clinical Microbiology and Infection: The Official Publication of the European Society of Clinical Microbiology and Infectious Diseases. 2004;**10**(4):272-288

[21] Naas T, Mikami Y, Imai T, Poirel L, Nordmann P. Characterization of In53, a class 1 plasmid- and composite transposon-located integron of *Escherichia coli* which carries an unusual array of gene cassettes. Journal of Bacteriology. 2001;**183**(1):235-249

[22] Su LH, Chen HL, Chia JH, Liu SY, Chu C, Wu TL, et al. Distribution of a transposon-like element carrying blaCMY-2 among *Salmonella* and other *Enterobacteriaceae*. The Journal of Antimicrobial Chemotherapy. 2006;**57**(3):424-429

[23] de Toro M, Fernández J, García V, Mora A, Blanco J, de la Cruz F, et al. Whole genome sequencing, molecular typing and in vivo virulence of OXA-48-producing *Escherichia coli* isolates including ST131 H30-Rx, H22 and H41 subclones. Scientific Reports. 2017;**7**(1):1-12

[24] Grindley ND. Transposition of Tn3 and related transposons. Cell. Jan, 1983;**32**(1):3-5

[25] Leff LG, Dana JR, McArthur JV, Shimkets LJ. Detection of Tn5-like sequences in kanamycin-resistant stream bacteria and environmental DNA. Applied and Environmental Microbiology. 1993;**59**(2):417-421

[26] Kraft CA, Timbury MC, Platt DJ. Distribution and genetic location of Tn7 in trimethoprim-resistant *Escherichia coli*. Journal of Medical Microbiology. 1986;**22**(2):125-131

[27] Alton NK, Vapnek D. Nucleotide sequence analysis of the chloramphenicol resistance transposon Tn9. Nature. 1979;**282**(5741):864-869

[28] Chalmers R, Sewitz S, Lipkow K, Crellin P. Complete nucleotide sequence of Tn10. Journal of Bacteriology. 2000;**182**(10):2970-2972

[29] Fernandez-Alarcon C, Singer RS, Johnson TJ. Comparative genomics of multidrug resistance-encoding IncA/C plasmids from commensal and pathogenic *Escherichia coli* from multiple animal sources. PLoS One. 2011;**6**(8):e23415

[30] Bozcal E, Dagdeviren M, Uzel A, Skurnik M. LuxCDE-luxAB-based promoter reporter system to monitor the *Yersinia enterocolitica* O:3 gene expression in vivo. PLoS One. 2017;**12**(2):e0172877

[31] Thomas CM. Plasmid Incompatibility. In: Bell E, editor. New York, NY: Springer; 2014

[32] Velappan N, Sblattero D, Chasteen L, Pavlik P, Bradbury AR. Plasmid incompatibility: More compatible than previously thought? Protein Engineering, Design and Selection. 2007;**20**(7):309-313

[33] Liakopoulos A, van der Goot J, Bossers A, Betts J, Brouwer MSM, Kant A, et al. Genomic and functional

characterisation of IncX3 plasmids encoding blaSHV-12 in *Escherichia coli* from human and animal origin. Scientific Reports. 2018;**8**(1):1-3

[34] Irrgang A, Hammerl JA, Falgenhauer L, Guiral E, Schmoger S, Imirzalioglu C, et al. Diversity of CTX-M-1-producing *E. coli* from German food samples and genetic diversity of the *blaCTX-M-1* region on IncI1 ST3 plasmids. Veterinary Microbiology. 2018;**221**:98-104

[35] Partridge SR, Kwong SM, Firth N, Jensen SO. Mobile genetic elements associated with antimicrobial resistance. Clinical Microbiology Reviews. 2018;**31**(4):e00088-17

[36] Koraimann G. Spread and persistence of virulence and antibiotic resistance genes: A ride on the F plasmid conjugation module. EcoSal Plus. 2018;**8**(1):1-23

[37] Matamoros S, van Hattem JM, Arcilla MS, Willemse N, Melles DC, Penders J, et al. Global phylogenetic analysis of *Escherichia coli* and plasmids carrying the mcr-1 gene indicates bacterial diversity but plasmid restriction. Scientific Reports. 2017;**7**(1):1-10

[38] Xavier BB, Lammens C, Ruhal R, Kumar-Singh S, Butaye P, Goossens H, et al. Identification of a novel plasmid-mediated colistin-resistance gene,mcr-2, in *Escherichia coli,* Belgium, June 2016. Eurosurveillance. 2016;**21**(27):1-6

[39] Yin W, Li H, Shen Y, Liu Z, Wang S, Shen Z, et al. Novel plasmid-mediated colistin resistance gene mcr-3 in *Escherichia coli*. MBio. 2017;**8**(3):e00543-17

[40] Rebelo AR, Bortolaia V, Kjeldgaard JS, Pedersen SK, Leekitcharoenphon P, Hansen IM, et al. Multiplex PCR for detection of plasmid-mediated colistin resistance determinants, mcr-1, mcr-2, mcr-3, mcr-4 and mcr-5 for surveillance purposes. Euro Surveillance. 2018;**23**(6):1-11

[41] Roer L, Overballe-Petersen S, Hansen F, Schønning K, Wang M, Røder BL, et al. *Escherichia coli* sequence type 410 is causing new international high-risk clones. Msphere. 2018;**3**(4):1-14

[42] Phornphisutthimas S, Thamchaipenet A, Panijpan B. Conjugation in *Escherichia coli:* A laboratory exercise. Biochemistry and Molecular Biology Education. 2007;**35**(6):440-445

[43] Norman A, Hansen LH, Sorensen SJ. Conjugative plasmids: Vessels of the communal gene pool. Philosophical Transactions of the Royal Society of London. Series B, Biological Sciences. 2009;**364**(1527):2275-2289

[44] Grohmann E, Muth G, Espinosa M. Conjugative plasmid transfer in gram-positive bacteria. Microbiology and Molecular Biology Reviews. 2003;**67**(2):277-301

[45] Zhang WJ, Wang XM, Dai L, Hua X, Dong Z, Schwarz S, et al. Novel conjugative plasmid from *Escherichia coli* of swine origin that coharbors the multiresistance gene cfr and the extended-spectrum-beta-lactamase gene blaCTX-M-14b. Antimicrobial Agents and Chemotherapy. 2015;**59**(2):1337-1340

[46] Moran RA, Holt KE, Hall RM. pCERC3 from a commensal ST95 *Escherichia coli*: A ColV virulence-multiresistance plasmid carrying a sul3-associated class 1 integron. Plasmid. 2016;**84-85**:11-19

[47] Wong MH, Xie M, Xie L, Lin D, Li R, Zhou Y, et al. Complete sequence of a F33:A-:B- conjugative plasmid carrying

the oqxAB, fosA3, and blaCTX-M-55 elements from a foodborne *Escherichia coli* strain. Frontiers in Microbiology. 2016;7:1729

[48] Brzuszkiewicz E, Thürmer A, Schuldes J, Leimbach A, Liesegang H, Meyer F-D, et al. Genome sequence analyses of two isolates from the recent *Escherichia coli* outbreak in Germany reveal the emergence of a new pathotype: Entero-aggregative-haemorrhagic *Escherichia coli* (EAHEC). Archives of Microbiology. 2011;**193**(12):883-891

[49] Ahmed N, Mellmann A, Harmsen D, Cummings CA, Zentz EB, Leopold SR, et al. Prospective genomic characterization of the German enterohemorrhagic *Escherichia coli* O104:H4 outbreak by rapid next generation sequencing technology. PLoS One. 2011;**6**(7)

[50] Kampmeier S, Berger M, Mellmann A, Karch H, Berger P. The 2011 German Enterohemorrhagic *Escherichia coli* O104:H4 outbreak—The danger is still out there. Current Topics in Microbiology and Immunology, 2018.

[51] Shridhar PB, Patel IR, Gangiredla J, Noll LW, Shi X, Bai J, et al. Genetic analysis of virulence potential of *Escherichia coli* O104 serotypes isolated from cattle feces using whole genome sequencing. Frontiers in Microbiology. 2018;**9**

[52] Verdet C, Gautier V, Chachaty E, Ronco E, Hidri N, Decre D, et al. Genetic context of plasmid-carried blaCMY-2-like genes in *Enterobacteriaceae*. Antimicrobial Agents and Chemotherapy. 2009;**53**(9):4002-4006

[53] Liu H, Zhu B, Liang B, Xu X, Qiu S, Jia L, et al. A novel mcr-1 variant carried by an IncI2-type plasmid identified from a multidrug resistant Enterotoxigenic *Escherichia coli*. Frontiers in Microbiology. 2018;**9**

[54] Boisen N, Hansen A-M, Melton-Celsa AR, Zangari T, Mortensen NP, Kaper JB, et al. The presence of the pAA plasmid in the German O104:H4 Shiga toxin type 2a (Stx2a)–producing Enteroaggregative *Escherichia coli* strain promotes the translocation of Stx2a across an epithelial cell monolayer. The Journal of Infectious Diseases. 2014;**210**(12):1909-1919

[55] Fierz L, Cernela N, Hauser E, Nüesch-Inderbinen M, Stephan R. Characteristics of Shigatoxin-producing *Escherichia coli* strains isolated during 2010-2014 from human infections in Switzerland. Frontiers in Microbiology. 2017;**8**

[56] Barr JJ, Auro R, Furlan M, Whiteson KL, Erb ML, Pogliano J, et al. Bacteriophage adhering to mucus provide a non-host-derived immunity. Proceedings of the National Academy of Sciences of the United States of America. 2013;**110**(26):10771-10776

[57] Colavecchio A, Cadieux B, Lo A, Goodridge LD. Bacteriophages contribute to the spread of antibiotic resistance genes among foodborne pathogens of the *Enterobacteriaceae* family—A review. Frontiers in Microbiology. 2017;**8**

[58] Penades JR, Chen J, Quiles-Puchalt N, Carpena N, Novick RP. Bacteriophage-mediated spread of bacterial virulence genes. Current Opinion in Microbiology. 2015;**23**:171-178

[59] Iyer A, Barbour E, Azhar E, Qadri I, Chaudhary A, Abuzenadah A, et al. Bacteriophages in *Escherichia coli* antimicrobial resistance. Advances in Bioscience and Biotechnology. 2013;**04**(03):469-476

[60] Marinus MG, Poteete AR. High efficiency generalized transduction in *Escherichia coli* O157:H7. F1000Research. 2013;**2**:7

[61] Marti E, Variatza E, Balcazar JL. Bacteriophages as a reservoir of extended-spectrum beta-lactamase and fluoroquinolone resistance genes in the environment. Clinical Microbiology and Infection. 2014;**20**(7):O456-O4O9

[62] Billard-Pomares T, Fouteau S, Jacquet ME, Roche D, Barbe V, Castellanos M, et al. Characterization of a P1-like bacteriophage carrying an SHV-2 extended-spectrum beta-lactamase from an *Escherichia coli* strain. Antimicrobial Agents and Chemotherapy. 2014;**58**(11):6550-6557

[63] Barondess JJ, Beckwith J. A bacterial virulence determinant encoded by lysogenic coliphage lambda. Nature. 1990;**346**(6287):871-874

[64] Vica Pacheco S, Garcia Gonzalez O, Paniagua Contreras GL. The Lom gene of bacteriophage lambda is involved in *Escherichia coli* K12 adhesion to human buccal epithelial cells. FEMS Microbiology Letters. 1997;**156**(1):129-132

[65] Vlot M, Nobrega FL, Wong CFA, Liu Y, Brouns SJJ. Complete genome sequence of the *Escherichia coli* phage Ayreon. Genome Announcements. 2018;**6**(2)

[66] Strauch E, Schaudinn C, Beutin L. First-time isolation and characterization of a bacteriophage encoding the Shiga toxin 2c variant, which is globally spread in strains of *Escherichia coli* O157. Infection and Immunity. 2004;**72**(12):7030-7039

[67] Amarillas L, Chaidez C, Gonzalez-Robles A, Lugo-Melchor Y, Leon-Felix J. Characterization of novel bacteriophage phiC119 capable of lysing multidrug-resistant Shiga toxin-producing *Escherichia coli* O157:H7. PeerJ. 2016;**4**:e2423

[68] Hernandez-Doria JD, Sperandio V. Bacteriophage transcription factor Cro regulates virulence gene expression in enterohemorrhagic *Escherichia coli*. Cell Host and Microbe. 2018;**23**(5): 607-17 e6

[69] Morita M, Tanji Y, Mizoguchi K, Akitsu T, Kijima N, Unno H. Characterization of a virulent bacteriophage specific for *Escherichia coli* O157:H7 and analysis of its cellular receptor and two tail fiber genes. FEMS Microbiology Letters. 2002;**211**(1):77-83

[70] Lloyd AL, Rasko DA, Mobley HL. Defining genomic islands and uropathogen-specific genes in uropathogenic *Escherichia coli*. Journal of Bacteriology. 2007;**189**(9):3532-3546

[71] Li J, Tai C, Deng Z, Zhong W, He Y, Ou HY. VRprofile: Gene-cluster-detection-based profiling of virulence and antibiotic resistance traits encoded within genome sequences of pathogenic bacteria. Briefings in Bioinformatics. 2017

[72] Mohammad DS. On the role of genomic Islands in bacterial pathogenicity and antimicrobial resistance. International Journal of Advanced Biotechnology and Bioengineering. 2014;**2**(1):18-36

[73] Nieto PA, Pardo-Roa C, Salazar-Echegarai FJ, Tobar HE, Coronado-Arrazola I, Riedel CA, et al. New insights about excisable pathogenicity islands in salmonella and their contribution to virulence. Microbes and Infection. 2016;**18**(5):302-309

[74] Blum G, Falbo V, Caprioli A, Hacker J. Gene clusters encoding the cytotoxic necrotizing factor type 1, Prs-fimbriae and alpha-hemolysin

form the pathogenicity island II of the uropathogenic *Escherichia coli* strain J96. FEMS Microbiology Letters. 1995;**126**(2):189-195

[75] Perna NT, Mayhew GF, Posfai G, Elliott S, Donnenberg MS, Kaper JB, et al. Molecular evolution of a pathogenicity island from enterohemorrhagic *Escherichia coli* O157:H7. Infection and Immunity. 1998;**66**(8):3810-3817

[76] Schubert S, Rakin A, Karch H, Carniel E, Heesemann J. Prevalence of the "high-pathogenicity island" of Yersinia species among *Escherichia coli* strains that are pathogenic to humans. Infection and Immunity. 1998;**66**(2):480-485

[77] Karch H, Schubert S, Zhang D, Zhang W, Schmidt H, Olschlager T, et al. A genomic island, termed high-pathogenicity island, is present in certain non-O157 Shiga toxin-producing *Escherichia coli* clonal lineages. Infection and Immunity. 1999;**67**(11):5994-6001

[78] Middendorf B, Blum-Oehler G, Dobrindt U, Muhldorfer I, Salge S, Hacker J. The pathogenicity islands (PAIs) of the uropathogenic *Escherichia coli* strain 536: Island probing of PAI II536. The Journal of Infectious Diseases. 2001;**183**(Suppl 1):S17-S20

[79] Clermont O, Bonacorsi S, Bingen E. The *Yersinia* high-pathogenicity island is highly predominant in virulence-associated phylogenetic groups of *Escherichia coli*. FEMS Microbiology Letters. 2001;**196**(2):153-157

[80] Schneider G, Dobrindt U, Bruggemann H, Nagy G, Janke B, Blum-Oehler G, et al. The pathogenicity island-associated K15 capsule determinant exhibits a novel genetic structure and correlates with virulence in uropathogenic *Escherichia coli* strain 536. Infection and Immunity. 2004;**72**(10):5993-6001

[81] Schubert S, Picard B, Gouriou S, Heesemann J, Denamur E. Yersinia high-pathogenicity island contributes to virulence in *Escherichia coli* causing extraintestinal infections. Infection and Immunity. 2002;**70**(9):5335-5337

[82] Parreira VR, Gyles CL. A novel pathogenicity island integrated adjacent to the thrW tRNA gene of avian pathogenic *Escherichia coli* encodes a vacuolating autotransporter toxin. Infection and Immunity. 2003;**71**(9):5087-5096

[83] Yang B, Wang S, Huang J, Yin Z, Jiang L, Hou W, et al. Transcriptional activator GmrA, encoded in Genomic Island OI-29, controls the motility of enterohemorrhagic *Escherichia coli* O157:H7. Frontiers in Microbiology. 2018;**9**:338

[84] Ho WS, Ou HY, Yeo CC, Thong KL. The dnd operon for DNA phosphorothioation modification system in *Escherichia coli* is located in diverse genomic islands. BMC Genomics. 2015;**16**:199

[85] Wang G, Zhou F, Olman V, Li F, Xu Y. Prediction of pathogenicity islands in enterohemorrhagic *Escherichia coli* O157:H7 using genomic barcodes. FEBS Letters. 2010;**584**(1):194-198

[86] Chouikha I, Germon P, Bree A, Gilot P, Moulin-Schouleur M, Schouler C. A selC-associated genomic island of the extraintestinal avian pathogenic *Escherichia coli* strain BEN2908 is involved in carbohydrate uptake and virulence. Journal of Bacteriology. 2006;**188**(3):977-987

Section 3

Preventing *Escherichia coli* Infections

Chapter 5

Escherichia coli and Food Safety

Gözde Ekici and Emek Dümen

Abstract

Foodborne pathogens are evaluated as an important risk factor in terms of public health in developed and developing countries due to their extensiveness all around the world. *Escherichia coli* and other coliform bacteria are important foodborne pathogens. Some of the most important sources of contamination for these groups of microorganisms are reported as: areas with unfavorable hygiene, contaminated waste water, meat products, cereal products and vegetables. Total coliform bacteria and *E. coli* count is known to be the indicator of unfavorable hygienic conditions and fecal contamination in foods. Foodborne diseases, however, are a global issue. A joint approach by all countries and related international organizations is a prerequisite for detection and control of foodborne problems that pose a threat to human health and international trade. Despite their complicated biology, epidemiology and analyses, most foodborne diseases are preventable. It is of vital importance for public health that consumers and food producers act in accordance with the principles regarding internationally accepted safety methods.

Keywords: *Escherichia coli*, food safety, foodborne diseases

1. Introduction

Microorganism of varying types and numbers can be found on food of animal and plant origin. The types and number of microorganism on food can be changed due to food processing, inappropriate purchasing, storing, preparing, cooking or serving. Increase in the number of these microorganisms due to the abovementioned changes may lead to spoiling of the food, causing a pathogenic effect on humans. The most important of foodborne pathogenic bacteria is *Escherichia coli*. It is transmitted through fecal or oral route and it should, under no circumstances, be present in any food. The most prominent symptom caused by this microorganism is its diarrheagenic effect. Moreover, it is known to cause sepsis, meningitis and many enteric diseases. Inability to ensure food safety is one of the biggest food-related problems. Food safety means ensuring necessary hygienic conditions and taking protective safety precautions for a healthy and safe food production throughout all processes from obtaining raw materials to production, transportation, storage, distribution and consumption of food. This section will focus on the pathogenic characteristics of food contaminated with *E. coli*, food contamination cases, current food safety approaches and methods of prevention/protection.

2. *Escherichia coli* and food poisoning

Escherichia coli, one of the 30 members of the bacterial family of *Enterobacteriaceae*, is a coliform bacterium and is one of the 6 types of *Escherichia* species (*E. adecaroxylate, E. blattae, E. fergusonii, E. hermannii* and *E. vulneris*). It is a gram-negative, non-spore-forming, facultative, anaerobic, rod shaped, mesophilic bacterium that grows in 7–45°C. The group of coliform bacteria consists of *Citrobacter, Enterobacter, Klebsiella* and *Escherichia*. While there are bacteria of fecal origin among coliform bacteria, there are also bacteria of plant origin such as *Enterobacter aerogenes, Citrobacter freundii*, and *Klebsiella pneumoniae*. Presence of coliform group in food is indicative of fecal contamination, poor hygienic conditions or existence of enteric pathogens. For instance, the presence of coliform bacteria in raw milk is an indication of poor hygiene in milking or storage conditions. The presence of coliform bacteria in raw or frozen fruits and vegetables is not important as *Enterobacter, Citrobacter* and *Klebsiella* are naturally present in the microbiota of plants. However, *E. coli* presence in fruits and vegetables is very important in terms of inadequate hygiene. *E. coli* is an important pathogen as it is an indicator of fecal contamination in foods and drinking water. Due to this characteristic, it is considered as an indicator bacterium in food safety and hygiene [1–3].

Being the prominent bacterium in the facultative anaerobic microbiota of the intestines, *E. coli* is widespread in stool and the environment. Some of its pathogenic strains both cause intoxication by creating toxins and cause gastroenteritis, pathologic kidney and brain damage by causing infection-type food poisoning through cellular increase. Some enterotoxin producing *E. coli* strains are divided into two groups as heat-stable and heat-labile. The heat-stable toxin is known as stable toxin (ST) and the heat-labile toxin is known as labile toxin (LT). Both toxins can be found together or separately. Moreover, pathogenic strains are also known to cause serious diseases such as diarrhea, peritonitis, mastitis, septicemia, pneumonia and neonatal meningitis. Among gram-negative bacilli, *E. coli* is the most widespread pathogen that causes meningitis especially in neonatal period. It has serious morbidity and mortality rates worldwide. The mortality rates in neonatal meningitis cases are reported to vary between 15–40% and 50% of the survivors are reported to continue their lives with neurological damage [2, 4]. Intestinal pathogenic *E. coli* are classified as shown in **Figure 1** at least six subgroups/pathotypes as enterotoxigenic *E. coli* (ETEC), enteropathogenic *E. coli* (EPEC), enteroaggregative *E. coli* (EAEC), diffusely-adherent *E. coli* (DAEC), enteroinvasive *E. coli* (EIEC) and enterohemorrhagic *E. coli* (EHEC). EHEC is also known as Shiga toxin (stx) producing *E. coli* (STEC) and verotoxin producing *E. coli* (VTEC) [5, 6].

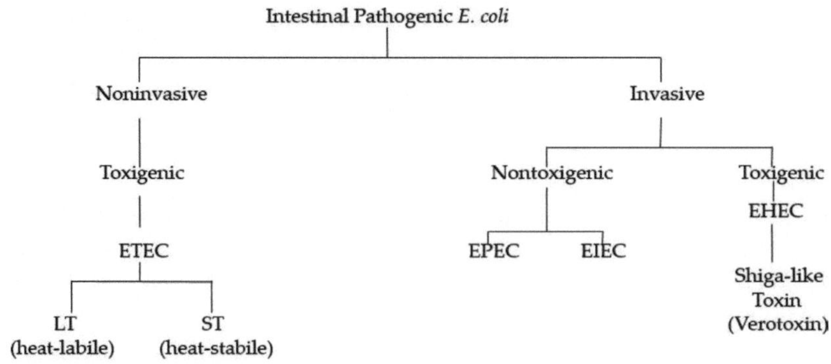

Figure 1.
Mechanism of intestinal pathogenic E. coli strains [6].

2.1 Enterotoxigenic *E. coli* (ETEC)

People living in developing countries have often been reported to have this pathotype in their feces and shown to have developed immunity against this microorganism. Being a cause of mortality in children under 5, the most frequently observed microorganism in childhood diarrhea is ETEC and it is also responsible for 30–60% of travelers' diarrhea. Infection is characterized by watery diarrhea and, depending on the person, its course may range from a normal course to cholera-like defecation with the addition of symptoms such as vomiting and high fever [2, 4, 7]. Diarrhea is the most common causes of mortality in society and among young children, especially those living in Asia and sub-Saharan Africa with inadequate healthcare systems and limited access to clean drinking water. Recent systematic studies have reported that each year an estimated 600,000 children under the age of 5 lose their lives. Diarrhea occurs due to the consumption of food or water contaminated with viral, bacterial or parasitic pathogens. Among these potential pathogens, the most common cause of diarrhea in children under five is the ETEC (heat-stable – ST and/or heat-labile – LT type toxin) producing *E. coli* strains. Through the production of fimbrial or non-fimbrial adhesins, ETEC strains cause hypersecretion of fluids by producing enterotoxins that disrupts fluid and electrolyte homeostasis in the epithelial cells of small intestines, leading to watery diarrhea. Without rehydration, moderate or severe diarrhea could lead to dehydration and acute mortality [7].

2.2 Enteropathogenic *E. coli* (EPEC)

It is known to be the oldest *E. coli* serotype causing diarrhea and its most important characteristic is adherence. In EPEC infections, vomiting and low body temperature are observed in addition to watery diarrhea [5]. It is known to cause diarrhea in infants and outbreaks can occur in neonatal care units. Humans, pigs and bovines may be infected with this microorganism. EPEC is transmitted from person to person, however; rarely, it is also known to spread through contaminated food and water [4, 6].

The ability to produce attaching and effacing (A/E) lesions is a distinctive phenotype for EPEC. Bacteria cause extensive deterioration on microvilli by strongly adhering to the host cell membrane. This adherence to the cell is mediated by an outer membrane protein called intimin. Moreover, depending on the presence of *E. coli* adherence factor – EAF), EPEC is classified as typical EPEC (tEPEC) and atypical (aEPEC) strains. In addition, as a distinctive factor, all EPEC strains lack the Shiga toxin (*stx*) producing genes. Among single-pathogen infections, EPEC has the second highest severity score after rotavirus, followed by ETEC. Diarrheagenic *E. coli*, especially EPEC, ETEC and EAEC are found out to be the main pathogens related to chronic diarrhea and its complications that lasts more than 14 days in developing countries. Moreover, among children with chronic diarrhea in developing countries, aEPEC was the most common pathogen isolated and it is the most common clinical case. These findings show that aEPEC may have a tendency to be naturally more chronic than other diarrheagenic *E. coli* [8, 9].

2.3 Enteroaggregative *E. coli* (EAEC)

This pathotype is a foodborne enteropathogen observed in acute and persistent diarrhea cases in children, patients with suppressed immune systems in developing countries and people traveling to endemic regions. Growth disorders and cognitive disorders in children living in developing countries, stem from EAEC infections. In the pathogenesis of EAEC, the first step is the strong adherence to the intestinal mucosa. The second step is leading to the development of enterotoxins and cytotoxins and the third step is known to be characterized with the ability to induce mucosal inflammation. Many different virulence factors regarding these three steps have been defined, however; none of them are present in all strains. Three adherence models related to EAEC have been defined. In addition to the localized adherence (LA) model that was defined first, there is also a diffuse adherence (DA) model and aggregative adherence (AA) model. The strains corresponding to the AA pattern were later defined as "Enteroadherent-aggregative *E. coli*". However, this term was then replaced with the current name "Enteroaggregative *E. coli*". AA phenotype has to be present in order for an *E. coli* strain of EAEC pathotype to be classified [10].

It is commonly found in foods in Mexico, including desserts and salsa sauces, and the visitors of the country are known to be more sensitive to EAEC infections during their stay rather than ETEC, which they are the most susceptible to. The reason behind this is the EAEC's ability to suppress the immune system and cause chronic infection. EAEC is also more resistant to antibiotics compared to the other diarrheagenic pathogens. Persistent infection and chronic disruption in intestinal functions cause malnutrition and decline in physical and mental development, especially in children. Malnutrition, which is observed due to micronutrient deficiency, induces infection. Development of infection induces malnutrition. This whole cycle increases the burden of acute diarrhea [11].

2.4 Diffusely-adherent *E. coli* (DAEC)

Hep-2 or HeLa cell cultures are called DAEC due to their diffuse adherence characteristics. DAEC serotypes are known to cause chronic diarrhea in children between the ages of 1 and 5. They cause degradation in the intestinal epithelium by binding to proteins that accelerate degradation. Mild diarrhea void of fecal leukocytes is the indication of infection. In France, DAEC strains were found out to be widespread in diarrhea cases observed in inpatients from a hospital with no other enteropathogen. This situation indicates that DAEC strains may be an important diarrheagenic pathogen in developed countries. Recent studies show that some DAEC strains contain virulence factors present in uropathogenic *E. coli* (UPEC) strains [5, 12].

2.5 Enteroinvasive *E. coli* (EIEC)

EIEC strains causing inflammatory damage in intestinal mucosa and submucosa are very similar to those produced by *Shigella*. These microorganisms have the same spreading and reproducing abilities inside epithelial cells. However, clinically, EIEC-related watery diarrhea is much more commonly observed than dysentery caused by *Shigella*. O antigens of EIEC can cross-react with O antigens of *Shigella*. The disease starts with severe abdominal cramping, weakness, watery stool, difficulty urinating and fever. It could rarely aggravate and turn into watery stool containing blood or mucus. The fecal leukocytes observed in shigellosis may also be observed in the mucus smear of a person infected with EIEC. EIEC infections are endemic to less developed countries and are reported to be rarely observed infections in developed countries. The incubation period is observed as 10–18 hours. There is evidence showing that EIEC is transmitted through contaminated foods. Just like in shigellosis, cases of diarrhea with enteroinvasive strains can be treated by using antimicrobials effective against *Shigella* isolates [13]. In a study conducted to investigate the effects of antibiotic usage, stool samples were analyzed to find out whether it affected pathogen findings. Four and fifty-six tourists from Finland were all informed about antibiotic usage during travel and stool samples were collected from them both before and after the travel. There were differences between the travelers that visited various countries before and the ones that did not use any antibiotics in terms of *Enterobacteriaceae* findings, as well as some health problems during the travel and pathogenic findings in stools [14].

2.6 Enterohemorrhagic *E. coli* (EHEC)

EHEC are also named Shiga toxin producing *E. coli* (STEC) and also verotoxin producing *E. coli* (VTEC). All strains of EHEC produce Shiga toxins that destroys vero cells similarly to Shiga toxins produced by Shigella. *E. coli* O157: H7, first defined after the outbreak associated with the consumption of rare cooked minced meat in 1982, is the primary cause of EHEC infection in industrialized countries including the USA, Canada and England. O26, O103, O111 and O145 can be listed among the other EHEC serogroups responsible for foodborne diseases. Even though the O157 strains are the ones that draw the most attention, the strains of other EHEC serogroups, especially O111, are gradually getting reported more and more around the world. Based on the severity of the disease, EHEC is regarded as the most serious *E. coli* strain among foodborne pathogens. *E. coli* O157:H7, differ from the other *E. coli* serotypes because of some of its characteristics, which are: not being able to grow in or above 42°C, not being able to ferment sorbitol, not having β-glucuronidase enzymes and producing enterohemolysins. Shiga-like toxin produced by *E. coli* O157:H7 is cytotoxic for human colon and duodenum. This toxin causes accumulation of fluid in intestines and lesions in colon through destruction of crypt epithelia. Intimin makes adhesion to the intestinal canal easier [5, 15].

EHEC has a wide spectrum including watery or bloody diarrhea and hemolytic uremic syndrome (HUS), which is an important factor in acute renal failure in children. The biggest EHEC O104:H4 outbreak was in Germany in 2011 with 855 HUS cases in 3842 people and 53 mortalities. This incidence, which raised concern all around the world, shows the importance of EHEC in terms of human health. Bovines are the main reservoir for these microorganisms to live on asymptomatically for years. Other smaller reservoirs for these microorganisms include sheep, goats, dogs, pigs and poultry. Other places where EHEC could stay alive for months include; bovine feces, soil and water. Butchering or processing of animals or contamination of plants through contaminated water or manure are the main routes for EHEC to spread to the food chain [16]. Following 3–12 days of incubation period after infection with *E. coli* O157:H7, watery diarrhea is observed as well as abdominal cramps and pain. In some cases, hemorrhagic colitis (HC) which is also known as bloody diarrhea, thrombotic thrombocytopenic purpura (TTP), fever and vomiting are included in the important clinical findings to be observed. Most patients recover within 10 days, however; depending on the serotype of the EHEC strain and *stx* subtype, HUS may develop 1 week after the start of diarrhea, that may lead to mortality especially in children and elderly people. HUS is characterized with acute renal failure, hemolytic anemia and thrombocytopenia. Coma, stroke, colon perforation, pancreatitis and hypertension are included among the other complications of HUS. It is estimated to lead to the early development of chronic renal failure in 15% of cases. Dialysis is necessary for HUS patients and mortality rate is 35%. Moreover, it is more commonly observed in women (70%) and during pregnancy (13%).Good treatment for this infection is still lacking, however; some new treatment strategies such as the usage of anti-vero toxin (anti-Shigatoxin) antibodies have been suggested. TTP, on the other hand, is clinically similar to HUS and fever, abdominal pain, gastrointestinal hemorrhage and central nervous system disorders are listed among complications that may develop. Frequently, it forms blood clots in the brain and result in mortality [2, 15–20].

The incidence and epidemiology of the important serotypes of *E. coli* are given in **Table 1**.

Pathogenic *E. coli*	Site of infection	Associated disease	Incidence	Target population	Significant transmission route
ETEC	Small intestine	Traveler's diarrhea, chronic childhood diarrhea (in developing countries)	16 U.S. outbreaks (1996–2003); prevalence 1.4% in patients with diarrhea; 79,420 cases of travelers' diarrhea each year (in the USA)	International travelers and children in developing countries	Food (raw produce, street vendors) and water
EPEC	Small intestine	Infant diarrhea	Hundreds of thousands of deaths world wide	Children in developing countries	Water, infant formula
EHEC	Large intestine	Hemorrhagic colitis (HC), hemolytic uremic syndrome (HUS)	110,000 cases and 61 deaths annually in the USA	All ages	Food (beef produce), person-to-person, water, animals
EIEC	Large intestine	Dysentery	Low in developed countries	Children in developing countries	Water (rare), person-to-person
EAEC	Intestine	Watery diarrhea with or without blood in the stool, acute and chronic	Developed and developing countries	Children and adults, travelers	Food, water, person-to-person

Table 1.
Summary of incidence and epidemiology of E. coli serotype [6].

3. Food safety and high-risk foods

Food safety means ensuring consumer safety and protecting products from biological, physical and chemical hazards throughout the whole process starting from the field to processing, storing, distributing, preparing and cooking [21]. In many countries around the world, people started to have a more conscious perspective on food and environment. Consumers tend to prefer food that is more natural, less processed, environment-friendly, healthy and produced safely. This tendency makes up the basis of the "preventive/protective" (pro-active) approach for measurements to be taken towards food safety both nationally and globally. This approach based on risk analysis is the most appropriate and effective method for controlling foodborne hazards. It also necessitates the application of proper control systems in the production chain [22]. Foodborne diseases are a global subject. A common approach by all countries and related international organizations is a prerequisite for the detection and control of foodborne problems threatening human health and international trade. Despite their complicated biology, epidemiology and analyses, most foodborne diseases are preventable. Public health institutions, food industry and consumers must be devoted to prevent foods from getting contaminated at farms, restaurants and homes. In outbreaks of foodborne diseases, continuous monitoring is vital for revealing the disease tendencies in foods, regions and associated pathogens. Genotype and subtype information obtained from contaminated strains are required for tracing the source of contamination, characterizing and comparing the strains [23].

The food safety management systems with a classical basis that were once accepted for safe production and consumption of foods has proven to be inefficient and researchers/organizations proposed the "risk-based food safety" approach. Risk-based food safety approach is significantly different than the classical hazard-based approach. In this regard, a food safety management system aims at estimating the risks to human health as well as defining, choosing and implementing strategies to control and decrease these risks. According to Codex Alimentarius, risk analysis is a process consisting of three components: risk assessment, risk management and risk communication. Today, the new approach is considered as an approach enabling food safety issues to be diagnosed more accurately and define strategies required to decrease these issues more effectively [23–25]. The principles of risk-based food safety are defined with a four-step framework. The first step includes a series of initial risk managements such as defining the food safety issues, developing a risk profile, setting risk management goals, deciding on the need for a risk assessment, forming a risk assessment policy, creating a risk assessment and/or risk ranking commission and analysis of the results following the assessment. In the second step, different risk management options are defined and the options are chosen after the assessment. The third step includes the implementation of risk management precautions. Lastly, in the last step, observations are carried out in appropriate areas within the food chain and this step is utilized in reviewing the effectiveness of the risk management precautions. This step usually includes public health monitoring in order to collect data on changes. In summary, this approach aims at improving the food safety in high-risk food/hazard combinations, decrease the burden of foodborne diseases and increase the consumer safety [25].

Billions of people in the world are under unsafe food risk. Each year, hundreds of thousands of people become sick or lose their lives due to consumption of unhygienic, high-risk foods. This is why safe food saves lives. In addition to improving the health of individuals and the public, safe food also boosts the economic growth in the regions where it is improved. Food safety covers four main areas, as shown in **Table 2** microbiological safety, chemical safety, personal hygiene and environmental hygiene [26].

Microbiological safety: the potential sources of foodborne diseases are bacterial agents. Diseases can range from mild gastroenteritis to neurological, hepatic or renal syndromes. Foodborne bacterial agents are primary cause of severe and fatal foodborne diseases. More than 90% of food poisoning diseases are caused by *Staphylococcus, Salmonella, Clostridium, Campylobacter, Listeria, Vibrio, Bacillus* and *E. coli* types.
Chemical safety: foods may contain some non-food chemical additives such as coloring agents or preservatives and contaminants such as pesticide residues. Heavy metals such as lead, cadmium, mercury and copper can be found in some food products possibly because of kitchen appliances or inadequate food hygiene.
Personal hygiene: inadequate personal hygiene in food processors or preparers can pose a great risk to public health. Simple activities such as hand washing and adequate washing facilities can prevent many foodborne diseases.
Environmental hygiene: inadequate or wrong recycling and lack of equipment for disposing of wastes lead to accumulation of spoiled and contaminated food. This situation than leads to the increase in the insect and bug populations contributing more to the risk of contamination and spoiling. For this reason, the hygienic conditions of the areas where food is processed and prepared are very important.

Table 2.
Four main areas of food safety [26].

4. Storage conditions and hygiene in foods

Controlling the entry of contaminants into the food chain can be difficult. In addition to poor hygiene, unfavorable transfer and storage conditions for foods or contaminated raw material usage also play a part in contamination. Low quality or contaminated foods may cause shipments to be canceled on an international level. This poses an obstacle for the trade between countries [27]. Food safety objectives are based on preventive actions such as safe raw material usage, good production practices and procedures with critical control points for hazard analysis. It is possible for the success of these preventive actions to reflect on the incidence of foodborne diseases. WHO and Center for Control of Foodborne Infections and Intoxications in Europe stated that one of the most important factors contributing to foodborne outbreaks were markers required for improving general hygiene and most of these were under the control of producers/consumers and listed these markers as following:

- Poor general hygiene
- Consuming raw products
- Using contaminated materials
- Contamination through infected people
- Cross-contamination
- Using contaminated tools
- Mistakes in processing
- Too early preparation
- Inadequate heating
- Inadequate warm-keeping
- Inadequate cooling
- Too long storage time
- Contamination during the last preparation phase
- Inadequate heating before reusing [28].

Attention should be paid to purchasing, preservation, preparation, cooking and serving processes for ensuring food hygiene and safety. While purchasing foodstuffs, attention should be paid to the shipment conditions, packaging and keeping the cold chain in potentially high-risk foods such as fish, meat, chicken and milk. Storage rules should be followed during storing. First in first out (FIFO) rule should be followed in storages. Temperature in storage units should be checked regularly and cooked meals should be left to cool down in room temperature before being stored in fridges. Shelves should be made of rustproof material and foods should be kept at least 15 cm away from the floor and walls.

Food	Preservation time (day)
Big piece of meat	3–5
Chicken	2–3
Minced meat	1–2
Sausage	2–3
Cooked meat	2–3
Raw fish	1–2
Shellfish	1
Cooked fish	2–3
Milk and cream	3–4
Eggs	14
Fruits	1–14
Vegetables	2–7

Table 3.
Preservation time for some foods [22].

There should be different sections for each food group (meat group, dairy group, fruit and vegetable group) so that cross-contamination is prevented. There is a risk of microorganism contamination from personnel, tools, environment or foods (cross-contamination) during the preparation phase. Color code system could be implemented in cutting areas to be able to prevent this from happening. Potentially high-risk foods should be processed without waiting. Cooked meals should be served in maximum 2 hours. Frozen foods should be thawed in 4–7°C. Internal temperature of poultry should be at least 75°C while cooking. Temperature of foods such as meat, fish and eggs should be increased to at least 63°C and they should be processed at this temperature for at least 2 minutes. Internal temperature of hot meals should be kept at 65°C in bain-marie with a closed lid. While serving food, clean containers should be used to transfer or hold the food. Cold foods should be kept under 4.5°C in a closed container. Preservation time is as important as preserving conditions when it comes to development, growing and spreading of microorganisms. Preservation times for some foods are listed in **Table 3** [22, 29].

5. Different pathotypes of *E. coli* and outbreaks

If we take a general look at the incidence and epidemiology of disease-causing *E. coli* pathotypes, we see many cases and outbreaks. For example; annual incidences of 31 primary pathogens were estimated in a study conducted in the USA in 2011. It is estimated that these 31 pathogens caused 6.6–12.7 million diseases; 39,500–75,700 hospitalizations and around 700–2300 mortalities. In another study conducted in the USA between 2003 and 2012, it was reported that foodborne outbreaks caused 4928 diseases, 1272 hospitalizations, 299 cases of HUS diagnosed by a physician and 33 deaths. The primary contamination sources were listed as 55% foodborne, 10% animal contact, 10% human-to-human transmission, 4% waterborne and 11% unknown reasons [28, 30]. In another study conducted in Argentina, O157:H7 STEC was detected 25.5% and non-O157 STEC was detected in 52.2% of the raw meats analyzed in terms of STEC. Argentina is one of the countries with the highest HUS incidence rates [31].

In meat products, non-O157 STEC prevalence varies between 2.4 and 30.0% for minced meat, 17.0 and 49.2% for sausage and 8.6 and 49.6 in meat put up for sale. When STEC contamination reports verifying that the STEC O157 prevalence had ranged between 0.2 and 27.8% for the last 30 years were assessed in terms of STEC O157 and non-O157 presence in bovine meat, non-O157 STEC rates were observed to be ranging between 2.1 and 70.1% [32]. Besides, EHEC serotypes were reported to stay alive for 9 months in −80°C and that they were not affected by the storage conditions of pieces of meat frozen in −20°C. *E. coli* is reported to be directly associated with consumption of undercooked meat. It is known that, especially meat and milk are very suitable environments for verotoxigenic *E. coli* and this microorganism produces significant amounts of verotoxins in contaminated meat kept in 37°C [33]. In studies, it was reported that the foods that are generally responsible for foodborne outbreaks were: meat and meat products, fish and seafood, chicken products, liver, ice cream, raw milk, rice meals, pasta and pasta salad, peanut, flour, cold sandwiches, fruit juices and raw fruits and vegetables [23]. In another study, it was stated that undercooked or uncooked hamburger, non-pasteurized fruit juices, raw vegetables contaminated with cow manure and infected cows are important sources of *E. coli*. For example; there were at least four deaths and over 500 laboratory approved infections were observed in an *E. coli* outbreak in 1993, related to hamburgers bought from a fast-food chain [6, 23]. The outbreak in the west of USA between 1992 and 1993 caused by *E. coli* O157:H7 that affected over 700 people and the outbreak in Japan in 1996 that affected over 8000 people and killed two people can be listed among the outbreaks caused by *E. coli* O157:H7. Foodborne outbreaks caused by *E. coli* O157:H7, O111:NM (non-motile) and STEC serotypes were reported in many countries such as Australia, Canada, Japan, USA, many European countries and North Africa [34].

In a study conducted on children's nursery in Japan between 2010 and 2013, it was detected that 68 of 1035 outbreaks were of EHEC origin. It is known that 30 of the 68 outbreaks (46%) were foodborne [35]. It is also known that there were two EIEC outbreaks reported in England in June of 2014. These cases are rare in England. However, it is emphasized that EIEC has a capacity to cause large and potentially serious gastrointestinal outbreaks in Europe and that it should be considered as a potential pathogen in foodborne outbreaks [36]. In 2011 (between May 1st and July 4th) 2971 STEC related gastroenteritis cases including 18 deaths and 845 HUS cases including 36 deaths were reported along with laboratory approval, among 3816 cases reported to the public health officials in Germany. Moreover, the number of HUS cases during outbreaks was reported to be approximately 70 times the figures that corresponds to the same period of previous years [37]. In another report from Germany, a case-control study was conducted with 26 patients with HUS and 81 control cases. The incidence of the disease was associated with kale consumption in the univariate analysis and with kale and cucumber consumption in the multivariate analysis. Twenty-five percent of the cases reported eating kale and 88% reported eating a salad [38]. In another case in Scotland in 1994, 71 cases were reported including 1 death and 11 HUS cases due to non-pasteurization of milk. In an *E. coli* O111 outbreak in Australia, 200 cases were reported including 23 HUS cases and 1 death due to a kind of sausage made from minced meat. The failure in chlorination of Municipality waters also caused outbreaks. In an outbreak in 2000 among campers in Aberdeenshire, 20 cases were reported due to the environmental exposure to the camp area contaminated by sheep manure. Among this group of campers, the number of people that the number of cases who had not washed their hands before meals was almost 9 times bigger than the number of people who became ill. It is a well-known fact that all *E. coli* outbreaks cause high costs for countries in addition to the severity of the infection and the damage it leaves on people [39].

6. Conclusion

Along the food chain, controllability and traceability are of great importance for ensuring the consumer safety and for foods to be protected from biological, physical and chemical hazards starting from the field to the moment of consumption. Consumers constitute the last ring of the food safety. The purchasing power and consciousness of consumers help ensure food safety and are the most important factors for protection and prevention against risks. Prevention of *E. coli* infections require not only developing new vaccines but also providing uncontaminated water and food. Food production companies should pay close attention to the cleanliness of their application areas and the disinfection of the running water. People who work in food facilities and services, should be given frequent trainings on hygiene so as to prevent *E. coli* contaminations. During travels from developed countries to developing countries, unsafe foods and foods that are sold out in the open should be avoided; packaged and labeled drinking water and beverages should be consumed [2, 40]. Cooking food at the right temperatures can ensure inactivation of *E. coli* as the factor is sensitive to high temperatures. While the meat is cut into pieces, the microorganisms on the surface of the meat reach the inner sections and can stay alive if a sufficient heat treatment is not applied, turning it into a risk factor for public health. Similarly, there are some potential risks in raw milk. It poses a risk if not pasteurized. *E. coli* can be inactivated with pasteurization [15, 41, 42].

Biological protection precautions are also very important. It is claimed that 8.0 $\log_{10}$ cfu/g lactic acid bacteria causes a 1.6 $\log_{10}$ cfu/g decrease in *E. coli* O157:H7; EHEC O157 multiplies by growing in the damaged Fuji apple, yet *Candida oleophila* may be effective in controlling this pathogen in these damaged apples. Decreasing this risk of contamination caused by farms, slaughterhouses, food producers and consumers is very important for protection from the pathogen *E. coli* O157:H7 strain, which can also be transmitted through food and water. Under the Food and Drug Administration (FDA), Meat Inspection Act and other regulations the food industry is responsible for producing safe foods that meet national standards, identify critical control points from production to consumption, and have good production practices. Hazard analysis and critical control points (HACCP) is a management system in which food safety is addressed through the analysis and control of biological, chemical and physical hazards for raw material production, procurement and packaging, distribution and consumption [2, 17, 40, 43].

Under the HACCP, the term hazard refers to any substance or condition that has the potential to cause adverse health effects and that is unacceptable. These hazards can be caused by the biological, chemical or physical contamination in the raw material, semi-processed or finished food product. Hazard analysis is defined as the assessment of the severity of the hazard and the likelihood of it happening. HACCP is a system managed based on seven principles to identify, assess and control possible hazards for food [17, 44];

1. Conduct hazard analysis

2. Identify critical control points (CCP)

3. Establish critical limits

4. Establish monitoring procedures

5. Establish corrective actions

6. Establish verification procedures

7. Establish documentation and record procedures [17].

These principles are accepted by state institutions, trade associations and the food industry. Today, food safety systems based on HACCP principles are successfully implemented in food processing facilities, retail food stores and global food service operations. Following HACCP rules in production facilities is vital. In a slaughterhouse in Mexico it was emphasized that HACCP should be applied in addition to antimicrobial treatment to reduce the presence of potential pathogens such as *E. coli* O157: H7 and non-O157 STEC in cattle carcasses. In the USA, it was stated that the most stringent measure for the prevention and control of EHEC is to determine the critical control points that lead to contamination of meat. Good manufacturing practices (GMP) and standard sanitation operating procedures (SSOPs) are accepted as the first steps in developing the HACCP system in the food industry. Successful implementation of GMPs and SSOPs is of great importance for the HACCP, because these systems are the building blocks of food safety during the processing phase [45]. These guidelines act as a guarantee for production, test, quality and assurance to help reduce the risk of foodborne diseases and ensure production and distribution of safe food for human consumption. Many countries follow GMP procedures and established their own GMP rules in accordance with their own legislations. The purpose of GMP is also help reduce the uncontrollable risks such as contamination and cross-contamination by testing the product. The main requirements for GMP are listed below [43, 45, 46];

- Comprehensible written instructions and procedures

- Trained employees

- Records of actions, mistakes and reviews

- Records of production and distribution

- Proper storage and distribution

- Complaint and recall systems [46].

In conclusion; it should not be forgotten that as a foodborne pathogen *E. coli* can spread in food, even in small numbers, and has the potential to cause infections, food poisoning and even death. Preventive measures include protecting the food from direct or indirect contamination, applying personal hygiene practices, preserving the processed food in appropriate places and temperatures, checking proper packaging and proper storage, cooking in proper temperatures, allowing proper cooling and keeping the cooked food away from raw food. There are many simple measures for consumers to take in order to prevent bacterial growth and ensure food safety. Consumers can develop their own safety methods at home by following the abovementioned measures. It is very important for food producers to comply with the safety method principles such as HACCP and GMP in terms of public health so as to prevent many diseases and outbreaks.

Author details

Gözde Ekici[1*] and Emek Dümen[2]

1 Department of Nutrition and Dietetics, Istanbul Kültür University, Istanbul, Turkey

2 Department of Food Hygiene and Technologies, Istanbul University – Cerrahpaşa, Istanbul, Turkey

*Address all correspondence to: g.ekici@iku.edu.tr

References

[1] Balpetek D. Examination of the presence of *E. coli* 0157:H7 in some meat products. Eurasian Journal of Veterinary Sciences. 2010;**26**(1):25-31

[2] Uçar G, Yörük NG, Güner A. *Escherichia coli* infections. Turkiye Klinikleri Journals Food Hygiene Technology. 2015;**1**(3):22-29

[3] Erkmen O. In: Erkmen O, editor. Microbiology of Food. 3rd ed. Ankara: Efil Press; 2013. 550 p

[4] Donnenberg MS. *Escherichia coli* Pathotypes and Principles of Pathogenesis. Baltimore, Maryland, USA: International Encyclopedia of Public Health; 2017. pp. 585-593

[5] Poole TL. In: Simjee S, editor. Foodborne Diseases. 1st ed. Totowa, New Jersey: Humana Press; 2007. 535 p

[6] Gerba CP. Environmentally transmitted pathogens. In: Environmental Microbiology. 3rd ed. Elsevier Inc; 2014. pp. 509-550

[7] Zhang W, Sack DA. Current progress in developing subunit vaccines against enterotoxigenic *Escherichia coli*-associated diarrhea. Clinical and Vaccine Immunology. 2015;**22**(9):983-991

[8] Ochoa TJ, Contreras CA. Enteropathogenic *E. coli* (EPEC) infection in children. Current Opinion in Infectious Diseases. 2011;**24**(5):478-483

[9] Katherine A, Sinfield R, Hart CA, Garner P. Pathogens associated with persistent diarrhoea in children in low and middle income countries: Systematic review. BMC Infectious Diseases;**9**(88)

[10] Elias WP, Navarro-Garcia F. Enteroaggregative *Escherichia coli* (EAEC). In: Torres AG, editor. *Escherichia coli* in the Americas. 1st ed. Galveston, TX, USA: Springer International Publishing; 2016. p. 384

[11] Okhuysen PC, DuPont HL. Enteroaggregative *Escherichia coli* (EAEC): A cause of acute and persistent diarrhea of worldwide importance. The Journal of Infectious Diseases. 2010;**202**(4):503-505

[12] Taddei CR, Moreno ACR, Filho AF, Montemor LPG, Martinez MB. Prevalence of secreted autotransporter toxin gene among diffusely adhering *Escherichia coli* isolated from stools of children. FEMS Microbiology Letters. 2003;**227**(2):249-253

[13] Brachman PS. In: Chin J, editor. Control of Communicable Diseases Manual. 17th ed. Vol. 154. Washington, USA: American Public Health Association; 2001. pp. 783-784

[14] Lääveri T, Vilkman K, Pakkanen S, Kirveskari J, Kantele A. Despite antibiotic treatment of travellers' diarrhoea, pathogens are found in stools from half of travellers at return. Travel Medicine and Infectious Disease. 2018;**23**(March):49-55

[15] Ertaş N, Yıldırım Y, Karadal F, Al S. The importance of *Escherichia coli* O157:H7 in foods of animal origin. Journal of The Faculty of Veterinary Medicine Erciyes University. 2013;**10**(1):45-52

[16] Karch H, Leopold SR, Kossow A, Mellmann A, Köck R, Bauwens A. In: Sing A, editor. Zoonoses Infections Affecting Humans and Animals. 1st ed. Germany: Springer Science Business Media; 2013. pp. 1689-1699

[17] Ibrahim OO. Foodborne Pathogen *Escherichia coli* O157:H7 History, Sources

of Transmission, Symptoms, Detection and Prevention. EC Microbiology. 2015;**2**(1):214-222

[18] Mele C, Noris M, Remuzzi G. Hemolytic uremic syndrome. Critical Care Nephrology. 2014;**36**(4):399-420

[19] Chapman PA, Ashton R. An evaluation of rapid methods for detecting *Escherichia coli* O157 on beef carcasses. International Journal of Food Microbiology. 2003;**87**(3):279-285

[20] Chapman PA, Ellin M, Ashton R, Shafique W. Comparison of culture, PCR and immunoassays for detecting *Escherichia coli* O157 following enrichment culture and immunomagnetic separation performed on naturally contaminated raw meat products. International Journal of Food Microbiology. 2001;**68**(1-2):11-20

[21] Onurlubaş E, Gürler AZ. The factors affecting level of consumers about food safety. Journal of Agricultural Faculty of Gaziosmanpasa University. 2016;**33**(1):132-141

[22] Seydi A. Risk factors and hygiene importance in food safety. Journal of Tourism Gastronomy Studies. 2017:310-321

[23] Bintsis T. Foodborne pathogens. AIMS Microbiology. 2017;**3**(3):529-563

[24] Koutsoumanis KP, Aspridou Z. Moving towards a risk-based food safety management. Current Opinion in Food Science. 2016;**12**:36-41

[25] Barlow SM, Boobis AR, Bridges J, Cockburn A, Dekant W, Hepburn P, et al. The role of hazard- and risk-based approaches in ensuring food safety. Trends in Food Science and Technology. 2015;**46**(2):176-188

[26] Fung F, Wang H, Menon S. Food safety in the 21st century. Biomedical Journal. 2018;**41**(2):88-95

[27] Korada SK, Yarla NS, Putta S, Hanumakonda AS, Lakkappa DB, Bishayee A, et al. A critical appraisal of different food safety and quality management tools to accomplish food safety. In: Food Safety and Preservation. Elsevier Inc; 2018. pp. 1-12

[28] Notermans SH. In: Batt CA, Tortorello M, editors. Encyclopedia of Food Microbiology. 2nd ed. Vol. 1. USA: Academic Press Elsevier; 2014. 3248 p

[29] Hacıoğlu N, Girgin GK. Evaluation of the HACCP system by the kitchen workers of hotels: A research in 5-star hotels. Business Administration Journal. 2008;**9**(2):281-301

[30] Heiman KE, Mody RK, Johnson SD, Griffin PM, Gould LH. *Escherichia coli* O157 outbreaks in the United States, 2003-2012. Emerging Infectious Diseases. 2015;**21**(8):1293-1301

[31] Castro VS, Carvalho RCT, Conte-Junior CA, Figuiredo EES. Shiga-toxin producing *Escherichia coli:* Pathogenicity, supershedding, diagnostic methods, occurrence, and foodborne outbreaks. Comprehensive Reviews in Food Science and Food Safety. 2017;**16**(6):1269-1280

[32] Hussein HS, Bollinger LM. Prevalence of *Shiga Toxin*-producing *Escherichia coli* in beef cattle. Journal of Food Protection. 2005;**68**(10):2224-2241

[33] Weeratna RD, Doyle MP. Detection and production of verotoxin 1 of *Escherichia coli* O157:H7 in food. Applied and Environmental Microbiology. 1991;**57**(10):2951-2955

[34] Güner A, Atasever M, Aydemir Atasever M. New emerging and re-emerging bacterial foodborne pathogens. Journal of the Faculty of Veterinary Medicine, Kafkas University. 2009;**18**(5):889-898

[35] Kanayama A, Yahata Y, Arima Y, Takahashi T, Saitoh T, Kanou K, et al. Enterohemorrhagic *Escherichia coli* outbreaks related to childcare facilities in Japan. BMC Infectious Diseases. 2015;**15**(1):2010-2013

[36] Newitt S, MacGregor V, Robbins V, Bayliss L, Anne Chattaway M, Dallman T, et al. Two linked enteroinvasive *Escherichia coli* outbreaks, Nottingham, UK, June 2014. Emerging Infectious Diseases. 2016;**22**(7):1178-1184

[37] Frank C, Werber D, Cramer JP, Askar M, Faber M, Heiden M, et al. Epidemic profile of Shiga-toxin-producing *Escherichia coli* O104:H4 outbreak in Germany. The New England Journal of Medicine. 2011;**365**(19):1771-1780

[38] Buchholz U, Bernard H, Werber D, Böhmer MM, Remschmidt C. German outbreak of *Escherichia coli* O104:H4 associated with sprouts. The New England Journal of Medicine. 2011;**365**(19):701-709

[39] Pennington H. *Escherichia coli* O157. Lancet. 2010;**376**(9750):1428-1435

[40] Hurd HS, Malladi S. An outcomes model to evaluate risks and benefits of *Escherichia coli* vaccination in beef cattle. Foodborne Pathogens and Disease. 2012;**9**(10):952-961

[41] Harewood P, Rippey S, Montesalvo M. Effect of gamma irradiation on shelf life and bacterial and viral loads in hard-shelled clams (*Mercenaria mercenaria*). Applied and Environmental Microbiology. 1994;**60**(7):2666-2670

[42] Clavero MR, Monk JD, Beuchat LR, Doyle MP, Brackett RE. Inactivation of *Escherichia coli* O157:H7, *Salmonella*, and *Campylobacter jejuni* in raw ground beef by gamma irradiation. Applied and Environmental Microbiology. 1994;**60**(6):2069-2075

[43] Cálix-Lara TF, Rajendran M, Talcott ST, Smith SB, Miller RK, Castillo A, et al. Inhibition of *Escherichia coli* O157: H7 and *Salmonella enterica* on spinach and identification of antimicrobial substances produced by a commercial lactic acid bacteria food safety intervention. Food Microbiology. 2014;**38**:192-200

[44] Tzouros NE, Arvanitoyannis IS. Implementation of hazard analysis critical control point (HACCP) system to the fish/seafood industry: A review. Food Review International. 2000;**16**(3):273-325

[45] Cusato S, Gameiro AH, Sant'Ana AS, Corassin CH, Cruz AG, Oliveira CAF. Assessing the costs involved in the implementation of GMP and HACCP in a small dairy factory. Quality Assurance and Safety of Crops & Foods. 2014;**6**(2):135-139

[46] Saerekui P. Good Manufacturing Practices. ISBT Science Series. 2009;**4**(531):6-10

Chapter 6

Human and Veterinary Vaccines against Pathogenic *Escherichia coli*

Mariano Larzábal, Angel A. Cataldi and Daniel A. Vilte

Abstract

Pathogenic *Escherichia coli* constitute an important current problem of public health and animal production. Efforts have been made to fight the infections caused by these bacteria, and in this chapter, we present the progress made up to date in the vaccines generated for this purpose. Different vaccines have been tested against the pathotypes responsible for human diseases such as diarrhea and urinary infections. Also, the poultry market has deserved the effort of the researchers to obtain a product that fights the *E. coli* strains that cause diseases in them. Finally, advances are also presented for the zoonotic enterohemorrhagic *E. coli* (EHEC), which are a different problem due to their low importance as a disease factor in cattle, but they are a very important pathogen in humans. In several of these fields, authorized products have been developed and are currently being marketed.

Keywords: pathogenic *Escherichia coli*, vaccine, human, cattle, virulence factors

1. Introduction

This chapter deals with the current developments on human and veterinary vaccines against pathogenic *Escherichia coli* of following pathotypes: enterohemorrhagic *E. coli* (EHEC) and Shiga toxin producing *E. coli* (STEC), enterotoxigenic *E. coli* (ETEC), extraintestinal pathogenic *E. coli* (ExPEC), in particular uropathogenic *E. coli* (UPEC), and avian pathogenic *E. coli* (APEC). Other pathotypes were not considered because of a lesser development related to vaccines. In some cases, only vaccines tested in the target species (human, cattle, chicken, etc.) were considered due to the high abundance of publications where experimental vaccines were tested on rodent or on other animal models.

2. Vaccines against EHEC/STEC

2.1 Vaccines against EHEC/STEC for humans

Different factors make the development of a vaccine difficult to prevent EHEC/STEC infection and hemolytic uremic syndrome (HUS) in humans. The lack of knowledge about what type of immune response may confer protection, and the multiplicity of infection routes comprising bovine-derived food products, leafy green vegetables, pool or drinking water, person-to-person transmission [1], and the lack of reliable animal models complicate the advance in this field.

Szu and Ahmed developed polysaccharide conjugate vaccines composed of detoxified lipopolysaccharide (LPS) from *E. coli* O157, covalently linked to a carrier protein and a recombinant exoprotein of *Pseudomonas aeruginosa* (rEPA) that has been used for conjugation of polysaccharides and proteins [2]. Phase I and Phase II clinical studies were conducted in adults and in children ranging from 2 to 5 years old, respectively [3]. The *E. coli* O157 conjugate vaccines were safe for all ages, and a positive humoral IgG response with bactericidal activity was found in both age populations. However, there were certain limitations for using LPS-based vaccines. For example, LPS failed to induce a long-lasting humoral immune response especially in children, and STEC non-O157 serotypes were not covered. In one attempt to compensate for this shortcoming, the same group conjugated O-polysaccharide with the B subunit of Shiga toxin (Stx1) [2]. However, this formulation did not neutralize Shiga toxin (Stx2), the toxin type most frequently found in severe HUS cases.

The main virulence factor of STEC/EHEC is the Shiga toxin (Stx); in consequence, it is an optimal target to elicit neutralizing antibodies. Subsequently, various Stx-based vaccine approaches have been attempted. A vaccine consisting of a poly-N-acetylglucosamine (PNAG, a surface polysaccharide of STEC) conjugated to the B subunit of Stx1 was produced. The antibodies raised in rabbit neutralized Stx1 potently, but modestly Stx2. Passive transfer of antibodies indicates that anti-PNAG could confer protection, but the cross-reacting neutralization of Stx2 is limited [4].

To date, no vaccines have been approved for human use, exposing a void in both treatment and prevention of EHEC O157:H7 infections. Vaccine research and development efforts have oriented to cattle as the main reservoir.

2.2 Vaccines against EHEC for cattle

Up to date, different vaccine compositions have been tested to reduce the colonization of the bovine and the environmental dissemination of EHEC O157:H7. These vaccines have different immunogenic, adjuvants, inoculation pathways, number of doses, and of course differ in their development and evaluation level in experimental and natural conditions. In this occasion, we decided to consider the proposals whose capacity of protection was evaluated in cattle.

The key factor for achieving a protective immune response in the animal is the immunogen. Looking for the available literature, we can observe that there are several candidates, mainly colonization factors, which we can classify in: type III secretion system (T3SS) components, siderophore receptors and porin proteins, bacterins, whole-cell envelopes, flagellin, Shiga toxins toxoids, attenuated *Salmonella*, and combinations between more than one of these.

2.2.1 Vaccines based on T3SS components

The components of the T3SS were the first to be used as vaccines, because it was already known for the essential role that proteins such as intimin, Tir, EspA, and EspB play in the adhesion of EHEC O157:H7 to the host cell [5–7]. In 2004, Potter et al. [8] tested a vaccine composed by a protein supernatant of EHEC O157:H7 (containing various Esps and Tir) with the adjuvant VSA3, in animals that were later challenged with *E. coli* O157:H7, as well as in animals in a clinical trial. They observed significant increase in serum antibodies against proteins of T3SS and O157 lipopolysaccharide. There was also a decrease in the number of bacteria in feces, in the number of shedder animals, and in the duration of excretion in the vaccinated group. The clinical trial showed a reduced prevalence of EHEC O157:H7 in typical feedlot conditions when cattle were vaccinated. In 2005, Van Donkersgoed et al. [9]

published a field trial in nine feedlots using a vaccine similar to Potter et al. [8], and they did not observe a significant association between vaccination and pen prevalence of fecal *E. coli* O157:H7. Probably, the differences in the preparation of the secreted proteins, in this case with formalin, a different adjuvant and a different vaccination strategy, could cause the failure. Later, this same preparation, without formalin treatment and with VSA3 adjuvant, was standardized and analyzed in studies in commercial feedlots of beef cattle with a two-dose regimen. The authors evaluated the probability to detect the microorganism from terminal rectal mucosa as a measure of gut colonization [10] and other large-scale clinical trials on commercially fed cattle to test the efficacy of the regimen to reduce the environmental transmission of EHEC O157:H7 [11]. They concluded that the two-dose vaccine regimen was effective to reduce the probability for *E. coli* O157:H7 colonization of the terminal rectum of cattle at slaughter and reduces the probability for environmental transmission of the bacteria within commercial cattle feeding systems [12]. This evidence was accompanied by the generation of a commercial product known as Econiche(TM), which was developed by the Canadian company Bioniche Life Sciences. The vaccine was approved in Canada and the United Kingdom [13, 14] and had a pending conditional license in the U.S. [15], but in 2014, the Bioniche Animal Health business was purchased by Vèntoquinol SA [16], and the production of the vaccine was discontinued.

On the other hand, there were other groups that evaluated recombinant factor of the T3SS in various combinations. Van Diemen et al. [17] evaluated the carboxy-terminal 280 amino acids of intimin γ and β alone or combined with the portions of Efa-1 (EHEC factor for adherence). Immunized calves induced antigen-specific serum IgG and, in some cases, salivary IgA responses, but did not reduce the magnitude or duration of excretion of EHEC O26:H- (intimin β) or EHEC O157:H7 (intimin γ) after an experimental challenge. Similarly, immunization of calves with the truncated Efa-1 protein did not protect against intestinal colonization by EHEC O157:H7.

The vaccination of calves with recombinant EspA by intramuscular and intranasal routes induced high titers of antigen-specific IgG and salivary IgA, but these responses did not protect calves from intestinal colonization after a challenge with *E. coli* O157:H7 [18].

In 2010, McNeilly et al. [19] assessed whether three purified proteins, intimin (C-terminal 531 amino acids), EspA, and Tir, could reduce shedding of EHEC O157:H7. Furthermore, they evaluated if the inclusion of purified H7 flagellin to the vaccine could modify the vaccination efficacy. They used the intramuscular route and the rectal submucosal route and obtained a significant increased response in serum anti-EspA, anti-intimin, and anti-Tir IgG. When H7 flagellin was present, mucosal IgA and IgG anti-H7 was generated. After experimental infection with EHEC O157:H7, cattle showed that immunization with these purified antigens could significantly reduce the total levels of bacterial excretion and that the addition of H7 flagellin can improve this effect. More recently [20], this group optimized the formulation of this vaccine and concluded that the immunization with a combination of EspA, intimin, and H7 flagellin causes a significant reduction in shedding of EHEC O157:H7, more enough to impact on transmission between animals.

Vilte et al. [21] evaluated a vaccine composed by the C-terminal 280 amino acids of intimin γ and EspB. The intramuscular immunization elicited significantly high levels of serum IgG antibodies. Antigen-specific IgA and IgG were also induced in saliva, but only the IgA response was significant. Following experimental challenge with *E. coli* O157:H7, a significant reduction in bacterial shedding, was observed in vaccinated calves.

2.2.2 Vaccines based on siderophor receptors (SRP) and porins proteins

This proposal is based on reducing the ability of the bacterium to obtain iron from the environment to decrease the level of infection [22]. Thornton et al. [23] assessed the efficacy of an SRP-composed vaccine (Epitopix LLC) to reduce the prevalence and fecal excretion of EHEC O157:H7 in calves after an experimental infection. A significant response in serum anti-SRP antibody titers was detected, and they concluded that the vaccination tended to decrease the fecal prevalence and concentration of EHEC O157:H7. In other study [24], this group evaluated the vaccine to control the burden of *E. coli* O157:H7 in feedlot cattle in field conditions. Vaccination with SRP was associated with the reduction of fecal concentration of EHEC O157:H7 and suggested to reduce the burden of these bacteria on cattle. In a third assay, the vaccine was evaluated in feedlot cattle naturally shedding *E. coli* O157. There were two different inoculum volumes of vaccine, 2 and 3 ml. They concluded that SRP vaccine at the 3 ml dose reduced prevalence of *E. coli* O157. These results led to the commercial elaboration of a product known as *E. coli* bacterial extract vaccine with SRP® technology [25] and manufactured by Pfizer Animal Health (Now Zoetis Services LLC). It has conditional license of the U.S. Department of Agriculture.

2.2.3 Vaccines based on bacterins and bacterial envelopes

To evaluate the protection conferred by a bacterin of EHEC O157:H7, van Diemen et al. [17] prepared a formalin-inactivated bacterin from EDL933nalR strain that was inoculated in a combined schedule by intramuscular (with Alu-Oil) and intranasal (mixed with cholera toxin B subunit) routes. It elicited significant IgG responses against intimin and LPS from *E. coli* O157:H7, but did not confer protection against intestinal colonization by EHEC O157:H7 after challenge.

In 2011, Sharma et al. [26] evaluated three heat-inactivated bacterins to reduce the fecal shedding of *E. coli* O157:H7. They used a *hha* + strain of *E. coli* O157:H7 and constructed a *hha* and *hha sepB* deletion mutants. These deletions enhance the expression and intracellular accumulation of T3SS proteins, respectively. There was a significant increase in IgG against LEE-encoded proteins in calves vaccinated with *hha* or *hha sepB* mutant bacterins compared to wild strain, and a reduction in the numbers of animals shedding EHEC O157:H7 and in the duration of the fecal shedding of bacteria in feces was also observed.

An alternative to bacterins was assayed by Vilte et al. [27] by means of empty envelopes of EHEC O157:H7 known as bacterial ghosts (BGs). These envelopes retain all surface components in a nondenatured form. Animals were vaccinated with BGs (without adjuvants) by subcutaneous route and elicited significant levels of specific IgG in serum. Following oral challenge with *E. coli* O157:H7, a significant reduction in both the duration and total bacterial shedding was observed in vaccinated calves.

2.2.4 Vaccines based on flagellin

In 2008, McNeilly et al. [28] assayed a systemic (intramuscular) and mucosal (intrarectal) immunization with purified H7 flagellin to evaluate its effects on the colonization of EHEC O157:H7 after a challenge. The vaccination induced high titers of anti-H7 IgG and IgA antibodies in both serum and nasal secretions by intramuscular injection, but the intrarectal route failed in generating any response against H7. With respect to colonization of EHEC O157:H7, they concluded that immunization reduced colonization rates and delayed peak shedding, but did not affect total bacterial fecal shedding.

2.2.5 Vaccines based on attenuated Salmonella

In 2010, Khare et al. [29] assessed a live attenuated recombinant *Salmonella enterica* serovar Dublin *aroA* expressing intimin. The recombinant *Salmonella* was inoculated three times by oral route, but this did not produce a significant increase of intimin-specific IgA in serum and feces. Interestingly, they observed a transient clearance of *E. coli* O157:H7 in feces from vaccinated calves that subsequently reduced colonization and shedding of bacteria after an experimental challenge.

2.2.6 Vaccines based on Shiga toxins

An attractive target to research in cattle constitutes the Shiga toxins (Stx), the more important virulence factor for human health. In fact, Stx modulates cellular immune responses in cattle [30–32]. For that, in 2018, Schmidt et al. [33] evaluated the response, in a calf cohort, to immunization with recombinant Shiga toxoids genetically inactivated (rStx1MUT/rStx2MUT). Calves were passively (colostrum from immunized cows) and actively (intramuscularly) vaccinated, and this generated a significant difference in serum antibody titers compared with a control group. There was no EHEC O157:H7 challenge, but the natural presence of fecal STEC was monitored, and they observed less fecal positive (by PCR) samples from calves vaccinated than those from control animals. It is interesting because this investigation was not restricted to a determined serotype of EHEC.

In other study, Martorelli et al. [34] combined recombinant intimin and EspB with the B subunit of Stx2 fused to *Brucella* lumazine synthase (BLS-Stx2B) in order to evaluate whether the presence of Stx was able to improve the effect of the vaccine on fecal shedding of EHEC O157:H7 following an experimental inoculation. The immunization generates antibodies against Stx2B in serum and intestinal mucosa, but a superior level of protection compared with the use of intimin and EspB alone was not observed.

As was seen, there were and there are numerous efforts looking for a solution to reduce the contamination of cattle and its environment for EHEC O157:H7 and other dangerous serotypes too. Even two commercial products have been achieved, one of which has unfortunately been removed from the market. However, the fact that this pathogen does not constitute a direct problem for farmers, and because EHEC are not a cause of severe illness in cattle, makes our work more challenging. We have not only to find an adequate immunogen or formulation or doses that have a good response, but it must also be attractive enough for farmers to take it as a possible and desirable alternative to collaborate with one health perspective.

3. Vaccines against ETEC

ETEC is one of the leading bacteria that causes 200 million diarrheal cases and between 170,000 and 380,000 deaths annually in the world [35, 36]. Children under 5 years of age in developing countries are the most affected by ETEC infections and 42,000 deaths have been reported only in 2013 [37]. As well, ETEC infections are the main cause of diarrhea reported in persons who travel to Latin America, Africa, and Asia [38], where approximately 10 million traveler's diarrhea cases have been reported worldwide per year [39, 40].

There have been several attempts to obtain a vaccine against ETEC. The greatest efforts have been focused on virulence factors such as fimbriae called colonization factor antigens (CFA) and colonization surface antigens (CS) and two enterotoxins, heat-labile (LT) and heat-stable (ST). These virulence factors are extremely important during the pathogenesis of ETEC. CFA promote the attachment to enterocytes in the small intestine and are critical for colonization. After the attachment, ETEC releases LT and/or ST enterotoxins that disrupt fluid and cause electrolyte homeostasis in small intestinal epithelial cells [41]. Therefore, a vaccine directed against CFA could prevent the adherence and intestinal colonization, avoiding the subsequent release of enterotoxins by ETEC. Although 23 immunologically distinct CFA adhesins have been identified, its high variation present in the different circulating strains worldwide has prevented the development of a protective vaccine [42–44]. Studies of killed whole-cell vaccines demonstrate the development of colonization factor antigen I (CFA/I) and LT IgA antibodies but only were protective against homologous strains [45, 46]. To date, isolated ETEC can be divided into 42 different clonal groups with a singular combination of colonization factors (CFs) and toxins [47]. Alternative approaches of CS targets have been evaluated. CFA/I fimbria, CS3, CS5, and CS6 are immunologically related to the more prevalent CFs covering a 50–80% of the clinical ETEC isolates. ACE527 and rCTB-CF are two whole-cell vaccines that include a wide repertory of CFs. Five CFA adhesins (CFA/I, CS2, CS3, CS5, and CS6), one CFA subunit (CS1), and the LT-B subunit compose the ACE527 vaccine, represented by three live attenuated ETEC strains [48, 49]. The orally inoculated ACE527 protects challenged adults with homologous strains [49, 50]; however, it had adverse effects on volunteers [51]. The rCTB-CF vaccine is composed by five formalin-killed ETEC strains, which presents CFA/I, CS1, CS2, CS3, CS4, and CS5 adhesins supplemented with recombinant B subunit of the cholera toxin (rCTB) [52, 53]. The immune response induced by rCTB-CF vaccine showed to reduce the risk of developing diarrhea in adult travelers [54], but presented little protection and some adverse effects in young children [55, 56]. Despite the improvements made to rCTB-CF and ACE527 [50, 51, 57], these vaccines fail to protect against some ETEC strains since they do not contain the heat-stable class a(STa) or LT-A antigens.

Neutralizing the effects of these enterotoxins is considered a highly effective approach for preventing ETEC diarrhea. However, the development of vaccines from toxoids has not presented satisfactory results either. Both LT and ST are potent toxins; therefore, no toxin can be used directly as a vaccine antigen. However, detoxified derivatives of LT including the B subunit (not toxic LT-B) have demonstrated immunological properties even as an adjuvant in many animal models [58–60]. The A subunit is also included in studies of ETEC LT (LT-A) vaccine. The purpose of this incorporation is to induce a mostly protective immune response [61, 62]. On the other hand, STa unlike LT is poorly immunogenic due to its small size.

Recent progress in toxoids antigens enhances the potential for developing an effective and safe subunit vaccine against ETEC diarrhea. A skin path vaccine containing LT toxin was applied to humans. Immunized adults developed strong IgG and IgA antibody responses to LT [63, 64], which reduced the incidence of moderate-to-severe diarrhea caused by ETEC in healthy adults traveling to Mexico or Guatemala [65]. A secondary study demonstrated that the LT patch provided protection against LT + ETEC diarrhea but provided no protection against STa + ETEC [66]. Therefore, the use of the LT patch alone cannot be considered a suitable approach for vaccinating against ETEC [67].

Subunit vaccine from a mutant LT toxin (mLT) has been proposed. Although it is safer than LT, up to now, mLT has not demonstrated a wide efficacy in the protection against diarrhea caused by ETEC [66]. However, it has been explored mainly as a vaccine adjuvant. mLT demonstrated a higher protective efficacy of vaccine candidates for whole cell ETEC and a CFA + candidate adhesin subunit vaccine [68]. Therefore, its function as adjuvant favors a greater response of the candidate as well as allows the generation of anti-LT response.

Most of the ETEC strains isolated from patients with diarrhea are STa+ alone or LT+. The low immunogenicity and the high need to generate an immune response against STa led the researcher to develop mLT-STa fusions. Results of mouse immunization studies showed that LT-STaN12S toxoid fusion induces neutralizing anti-STa antibodies [69]. The high titer in mice presented against both toxoids makes it a promising antitoxin subunit vaccine.

Alternative adhesion tip of the CfaE and multiepitope fusion antigen (MEFA) were used as a conservative antigen for the development of a broadly protective ETEC antiadhesin vaccine [70]. Nonhuman primate immunized with CfaE showed protection against a CFA/I ETEC challenge [71]. However, the coadministration of CfaE and mLT did not protect against ETEC strains expressing Sta. MEFA is represented by epitopes from the seven most important CFA adhesins expressed by ETEC strains which was strongly immunogenic inducing high titers of antibodies specific to all adhesins [72]. This combination is an efficient means of developing a vaccine for antigenically heterogeneous pathogens like ETEC.

Novel antigens, such as the glycoprotein EtpA and the outer membrane adhesin EaeH, have been identified by genome sequencing [73]. Antibodies against EtpA demonstrated a significant reduction in the colonization of mice by the challenge ETEC strain (H10407) [74]. The identification of new antigens could be the way to incorporate epitopes that allow a greater range of protection against the different ETEC strains. These new epitopes, incorporated into the candidate vaccines that contain the most conserved and representative virulence factors of ETEC, could enhance the protection against diarrhea caused by ETEC.

ETEC is the most common cause of *E. coli* diarrhea in farm animals, and in the first four days of calves, life can be responsible for severe diarrhea with high mortality [75]. The strains are characterized by the surface adhesins fimbriae being F5, F7, and F17, more frequently involved in diarrhea in calves [76–79]. In addition, CS31 adhesin is prevalent on isolates from calves with *E. coli* septicemia [80, 81]. In regards to toxins, STa is the only toxin associated with disease in neonatal calves infected with ETEC [82], rarely LT are identified [76, 83]. Killed ETEC possessing F5-fimbriae or purified F5 fimbriae are contained in the commercial vaccines for calves. These vaccines do not contain F17, CS31, or STa; however, the impact of their absence is unknown. The maternal vaccination with these vaccines protects the neonatal ETEC infections by passive colostral and lactogenic immunity [84, 85]. Once the lactation stage is over, the cattle being more resistant [86]. In this way, vaccination dams are an effective strategy to prevent ETEC diarrhea in neonates calves [87, 88].

4. Vaccines against ExPEC

ExPEC causes a vast majority of urinary tract infections (UTIs), mostly in women with highly common recurrent episodes. ExPEC pathotypes causing UTI are called uropathogenic *E. coli* (UPEC). A recent review of Nesta and Pizza describes progresses in UPEC vaccines [89]. Most of the vaccines are aimed to stimulate the mucosal immune system. Initial attempts to the development of vaccines against ExPEC infections have been unsuccessful [90, 91]. The immunogen in these vaccine was single-purified virulence factors such as hemolysin [92], pilin, or the O-specific polysaccharide LPS, conjugated to either *Pseudomonas aeruginosa* endotoxin A (TA) or cholera toxin (CT) as carrier proteins [93, 94]. Because of high heterogeneity of O-specific polysaccharide, the design of a polysaccharide vaccine able to prevent ExPEC infections has been extremely challenging [95]. The O18-polysaccharide conjugated to either cholera toxin or to *P. aeruginosa* exoprotein A (EPA) was safe and able to induce antibodies with opsonophagocytic killing activity (OPK) in human volunteers. IgG purified from immunized individuals were protective in mice in an *E. coli* O18 challenge sepsis model [93]. However, a further test with a 12-valent O-antigen showed difficulties of cross protection.

Three vaccines against UTI reached market status in different countries. Vaccines based on whole or lysed fractions of inactivated *E. coli* have been evaluated in human clinical trials and have been so far the most effective in inducing some degree of protection in patients with recurrent urinary tract infections. The sublingual vaccine Uromune, an inactivated whole preparation of *E. coli*, *Klebsiella pneumoniae*, *Proteus vulgaris*, and *Enterococcus faecalis*, evaluated as prophylactic treatment in a multicenter retrospective observational study, demonstrated a certain degree of clinical benefit in terms of reduced recurrence rate in women suffering recurrent UTI [96].

The Solco Urovac vaccine, a vaginal suppository polymicrobial vaccine consisting of 10 inactivated uropathogenic bacteria, including six *E. coli* serotypes, *Proteus mirabilis*, *Morganella morganii*, *K. pneumoniae*, and *E. faecalis* strains, showed a minimal efficacy in Phase I and two Phase II trials in women suffering of recurrent UTIs [97–99]. However, in two additional clinical studies, the vaginal mucosal vaccine given for a 14-week period increased the time to reinfection in UTI susceptible women, representing a valuable alternative to the antibiotic-based prophylactic regimens [98, 100].

One of the first vaccine tested was based on *E. coli* extract was presented by Frey et al. [101]. This development lead to Uro-Vaxom, a commercial vaccine that was assessed in larger clinical trials a few years later [102] leading to the recommendation of Uro-Vaxom for prophylactic treatment of patients with recurrent urinary tract infections. OM-89/Uro-Vaxom vaccine demonstrated modest protection in women [103]. However, in a more recent trial on 451 female subjects, the lyophilized lysate of 18 *E. coli* strains, OM-89/Uro-Vaxom, manufactured using a modified lytic process, based on alkaline chemical lysis and autolysis, failed to show a preventive effect on recurrent uncomplicated UTIs [104].

Other vaccines reached clinical trial status. The development of ExPEC4V, a novel tetravalent bioconjugate vaccine developed by Glaxo Smith Kline against extraintestinal pathogenic *E. coli*, started by an epidemiological screening of the prevalent *E. coli* serotypes causing infection in women in Switzerland, Germany, and the USA. The authors selected the O antigens from LPS from the prevalent serotypes. It was evaluated for safety, immunogenicity, and clinical efficacy in placebo-controlled phase Ib trial [105]. By glycoengineering, the O antigens were conjugated in *E. coli*. The vaccine was well tolerated and elicited a robust antibody response in patients suffering from recurrent UTIs. Data indicated a reduced incidence of UTIs after vaccination, especially for higher bacterial loads. Clinical trial was performed in a population of healthy women with a history of recurrent UTI allowed for an additional, preliminary assessment of the candidate's clinical efficacy. In a multicenter Phase Ib clinical trial, 92 healthy adult women with a history of recurrent UTI received a single injection of either intramuscular ExPEC4V or placebo. The authors concluded that the tetravalent *E. coli* bioconjugate vaccine candidate was well tolerated and elicited functional antibody responses against all vaccine serotypes [106].

Mobley et al. investigated four defined antigens (IreA, Hma, IutA, and FyuA) associated with iron uptake, as an immunogen to prevent UTI [107]. The adjuvant used was cholera toxin. They tested the formulation in mice and observed antigen-specific IgG response. High antibody titers correlate with low colony forming units (CFUs) of UPEC following transurethral challenge of vaccinated mice. In addition, sera from women with and without histories of UTI have been tested for antibody levels to vaccine antigens. They indicated that iron uptake components are a suitable target for vaccination against UTI. Later, it was observed that the iron receptor FyuA is present in 77% and it is highly conserved among UPEC isolates [108]. FyuA immunization of mice reduced the colonization of UPEC in bladder and kidney. Adhesins and bacterial appendages as flagella have a long history as immunogenic single antigens component of experimental vaccines against UTI. FliC (or pilin) and FimH (from type 1 fimbriae) were administered to mice as a fusion or mixed and elicited higher levels of serum and mucosal. Different combinations and adjuvants elicited good protection against UPEC [109].

5. Vaccines against APEC

APEC that belongs to the ExPEC pathotype is a major causative agent of colibacillosis, aerosacculitis, polyserositis, septicaemia, and other diseases in chickens, turkeys, and other avian species. It is responsible for significant loss for the poultry industry. Main APEC serogroups associated with disease are O1, O2, and O78.

An ideal vaccine for poultry has to be able to induce cross protection against various APEC serogroups capable of causing disease. To be deliverable via a massive immunization method such as administering the antigens in drinking water or feed, *in ovo* and spray, in order to immunize thousands of broiler chickens, must be used. And, the vaccine has to be administered at a young age so that the birds develop a protective immune response by the age of 21 days when they are most vulnerable to APEC infection [110].

Inactivated bacterin vaccines or autovaccines of APEC are frequently used in the field, but their protective efficacy was not demonstrated. Landman and van Eck studied the protection conferred in laying hens against *E. coli* peritonitis syndrome (EPS) disease. Vaccines were formulated either as aqueous suspension or as water-in-oil induced protection against homologous challenge, while protection against heterologous challenge was inconclusive. However, other study [111] indicated no protection against a challenge with homologous or heterologous strain, in spite of a raise of IgY titer in vaccinated animals.

A recombinant *Salmonella enterica* serovar *Typhimurium* strains expressing the heterologous O polysaccharide of *E. coli* O1 and O2 was used to immunize chickens and elicited production of serum IgG and mucosal sIgA antibodies against the LPS of APEC O1 and O2. The immune response induced resulted protective against a lethal dose of both APEC serogroup strains [112]. An attenuated *Salmonella (Δlon, ΔcpxR*, and *ΔasdA16*) delivery system containing the genes encoding P-fimbriae (*papA* and *papG*), aerobactin receptor (*iutA*), and CS31A surface antigen (*clpG*) of APEC was constructed, and its potential as a vaccine candidate against APEC infection in chickens was evaluated. It induced an immune response and an effective protection against colibacillosis caused by APEC [113].

Mixed recombinant APEC surface proteins EtsC (a type I secretion system protein), the porins OmpA and OmpT, and TraT of APEC were used as antigens to immunize chickens seeking for a broad protection against several serotypes of APEC. The experimental vaccine elicited specific IgY and the induction of diverse cytokines in spleen and resulted in a reduction of lesion scores in different organs and a reduction of bacterial loads in blood and organs [114].

A commercial vaccine (Gall N tect CBL) against avian colibacillosis for layer hens is produced and marketed in Japan since 2012. It consists of a live attenuated O78 APEC with a *Δcrp* deletion. A big trial in layer hens [115, 116] demonstrated that it prevents avian colibacillosis infection and improves productivity. Live attenuated APEC strains were used as experimental vaccines for various research groups in colibacillosis fields. Strains deleted in *aroA* [117], *carAB* [118], and *galE* [119] were tested. Another commercial vaccine, based in subunit components, is Nobilis (MSD) composed by F11-and FT-antigens of APEC in a water-in-oil emulsion. No trials have reported by the company, but Gregersen et al. in 2010 [120] observed that in a controlled trial the vaccine application did not affect the overall mortality rate between the vaccinated and control flocks, but mortality due to *E. coli* infections made up only 8.2% in vaccinated birds compared with 24.6% in unvaccinated birds. Also, differences in average first week mortality, average weight at 38 days, and food conversion rate among vaccinated and control birds, respectively, were not found.

6. Conclusion

A high interest in the development of vaccines against pathogenic *E. coli* occurred in recent years. This interest is related both to pathotypes affecting human and animal health. Few vaccines have been licensed and reached market and public health status. There is an intrinsic difficulty in directing the immune response to a bacterial species that is commonly part of the animal microbiota. The state of the art consists in identifying antigenic components that are exclusive of pathogenic subtypes.

In spite of these difficulties, science has gained a relevant knowledge of virulence, pathogenicity, genomics, and epidemiology of pathogenic *E. coli,* and with no doubt this will benefit vaccinology concerning pathogenic *E. coli*.

Conflict of interest

The authors declare no conflict of interest.

Author details

Mariano Larzábal[1], Angel A. Cataldi[1*] and Daniel A. Vilte[2]

1 Institute of Biotechnology, IABIMO-CONICET, Center of Veterinary and Agronomical Research, National Institution of Agricultural Technology, Argentina

2 Institute of Pathobiology, IPVet-CONICET, Center of Veterinary and Agronomical Research, National Institution of Agricultural Technology, Argentina

*Address all correspondence to: cataldi.angeladrian@inta.gob.ar

References

[1] Rivas M, Chinen I, Miliwebsky E, Masana M. Risk factors for Shiga toxin-producing *Escherichia coli*-associated human diseases. Microbiology Spectrum. 2014;**2**. DOI: 10.1128/microbiolspec.EHEC-0002-2013

[2] Szu SC, Ahmed A. Clinical studies of *Escherichia coli* O157:H7 conjugate vaccines in adults and young children. Microbiology Spectrum. 2014;**2**. DOI: 10.1128/microbiolspec.EHEC-0016-2013

[3] Ahmed A, Li J, Shiloach Y, Robbins JB, Szu SC. Safety and Immunogenicity of *Escherichia coli* O157 O-specific polysaccharide conjugate vaccine in 2-5-year-old children. The Journal of Infectious Diseases. 2006;**193**:515-521. DOI: 10.1086/499821

[4] Lu X, Skurnik D, Pozzi C, Roux D, Cywes-Bentley C, Ritchie JM, et al. A poly-N-acetylglucosamine-Shiga toxin broad-spectrum conjugate vaccine for Shiga toxin-producing *Escherichia coli*. MBio. 2014;**5**:e00974-e00914. DOI: 10.1128/mBio.00974-14

[5] DeVinney R, Stein M, Reinscheid D, Abe A, Ruschkowski S, Finlay BB. Enterohemorrhagic *Escherichia coli* O157:H7 produces Tir, which is translocated to the host cell membrane but is not tyrosine phosphorylated. Infection and Immunity. 1999;**67**:2389-2398

[6] Tzipori S, Gunzer F, Donnenberg MS, de Montigny L, Kaper JB, Donohue-Rolfe A. The role of the eaeA gene in diarrhea and neurological complications in a gnotobiotic piglet model of enterohemorrhagic *Escherichia coli* infection. Infection and Immunity. 1995;**63**:3621-3627

[7] Dean-Nystrom EA, Bosworth BT, Moon HW, O'Brien AD. *Escherichia coli* O157:H7 requires intimin for enteropathogenicity in calves. Infection and Immunity. 1998;**66**:4560-4563

[8] Potter AA, Klashinsky S, Li Y, Frey E, Townsend H, Rogan D, et al. Decreased shedding of *Escherichia coli* O157:H7 by cattle following vaccination with type III secreted proteins. Vaccine. 2004;**22**:362-369. DOI: 10.1016/j.vaccine.2003.08.007

[9] Van Donkersgoed J, Hancock D, Rogan D, Potter AA. *Escherichia coli* O157:H7 vaccine field trial in 9 feedlots in Alberta and Saskatchewan. The Canadian Veterinary Journal. 2005;**46**:724-728

[10] Smith DR, Moxley RA, Peterson RE, Klopfenstein TJ, Erickson GE, Bretschneider G, et al. A two-dose regimen of a vaccine against type III secreted proteins reduced *Escherichia coli* in beef cattle in commercial feedlots. Foodborne Pathogens and Disease. 2009;**6**:155-161

[11] Smith D, Moxley R, Peterson R, Klopfenstein T, Erickson G, Clowser S. A two-dose regimen of a vaccine against *Escherichia coli* O157:H7 type III secreted proteins reduced environmental transmission of the agent in a large-scale commercial beef feedlot clinical trial. Foodborne Pathogens and Disease. 2008;**5**:589-598. DOI: 10.1089/fpd.2008.0080

[12] Smith DR, Moxley RA, Klopfenstein TJ, Erickson GE. A randomized longitudinal trial to test the effect of regional vaccination within a cattle feedyard on *Escherichia coli* O157:H7 rectal colonization, fecal shedding, and hide contamination. Foodborne Pathogens and Disease. 2009;**6**:885-892. DOI: 10.1089/fpd.2009.0299

[13] *E. coli* Vaccine Gets Full CFIA License | Canadian Cattlemen n.d. https://www.canadiancattlemen.ca/

daily/e-coli-vaccine-gets-full-cfia-license [Accessed: October 8, 2018]

[14] UK Approves *E. coli* O157:H7 Vaccine for Cattle | Food Safety News n.d. https://www.foodsafetynews.com/2012/08/uk-approves-e-coli-o157h7-vaccine-for-cattle/ [Accessed: October 8, 2018]

[15] Econiche(TM) Vaccine Efficacy Challenge Study Summarized in The Canadian Journal of Veterinary Research n.d. https://www.prnewswire.com/news-releases/econichetm-vaccine-efficacy-challenge-study-summarized-in-the-canadian-journal-of-veterinary-research-121343543.html [Accessed: October 8, 2018]

[16] Vétoquinol to Acquire the Animal Health Business of Bioniche Life Sciences inc. | Vetoquinol n.d. https://www.vetoquinol.com/eng/content/vetoquinol-acquisition-animal-health-business-bioniche-life-sciences [Accessed: October 8, 2018]

[17] van Diemen P, Dziva F, Abu-Median A, Wallis T, van den Bosch H, Dougan G, et al. Subunit vaccines based on intimin and Efa-1 polypeptides induce humoral immunity in cattle but do not protect against intestinal colonisation by enterohaemorrhagic *Escherichia coli* O157:H7 or O26:H. Veterinary Immunology and Immunopathology. 2007;**116**:47-58. DOI: 10.1016/j.vetimm.2006.12.009

[18] Dziva F, Vlisidou I, Crepin VF, Wallis TS, Frankel G, Stevens MP. Vaccination of calves with EspA, a key colonisation factor of *Escherichia coli* O157:H7, induces antigen-specific humoral responses but does not confer protection against intestinal colonisation. Veterinary Microbiology. 2007;**123**:254-261. DOI: 10.1016/j.vetmic.2007.02.016

[19] McNeilly TN, Mitchell MC, Rosser T, McAteer S, Low JC, Smith DGE, et al. Immunization of cattle with a combination of purified intimin-531, EspA and Tir significantly reduces shedding of *Escherichia coli* O157:H7 following oral challenge. Vaccine. 2010;**28**:1422-1428. DOI: 10.1016/j.vaccine.2009.10.076

[20] McNeilly TN, Mitchell MC, Corbishley A, Nath M, Simmonds H, McAteer SP, et al. Optimizing the protection of cattle against *Escherichia coli* O157:H7 colonization through immunization with different combinations of H7 flagellin, Tir, intimin-531 or EspA. PLoS One. 2015;**10**:e0128391

[21] Vilte DA, Larzábal M, Garbaccio S, Gammella M, Rabinovitz BC, Elizondo AM, et al. Reduced faecal shedding of *Escherichia coli* O157:H7 in cattle following systemic vaccination with γ-intimin C280 and EspB proteins. Vaccine. 2011;**29**:3962-3968. DOI: 10.1016/j.vaccine.2011.03.079

[22] Emery D, Straub D, Huisinga R, Carlson B. Active immunization using a siderophore receptor protein. 6,027,736; 2000

[23] Thornton A, Thomson D, Loneragan G, Fox J, Burkhardt D, Emery D, et al. Effects of a siderophore receptor and porin proteins-based vaccination on fecal shedding of *Escherichia coli* O157:H7 in experimentally inoculated cattle. Journal of Food Protection. 2009;**72**:866-869

[24] Thomson DU, Loneragan GH, Thornton AB, Lechtenberg KF, Emery DA, Burkhardt DT, et al. Use of a siderophore receptor and porin proteins-based vaccine to control the burden of *Escherichia coli* O157:H7 in feedlot cattle. Foodborne Pathogens and Disease. 2009;**6**:871-877. DOI: 10.1089/fpd.2009.0290

[25] *E. coli* Bacterial Extract vaccine with SRP® technology. n.d. https://

www.zoetisus.com/_locale-assets/mcm-portal-assets/services/documents/srpecoli/srp_ecoli_conditional_license_study_tech_bulletin_final.pdf

[26] Sharma VK, Dean-Nystrom EA, Casey TA. Evaluation of *hha* and *hha sepB* mutant strains of *Escherichia coli* O157:H7 as bacterins for reducing *E. coli* O157:H7 shedding in cattle. Vaccine. 2011;**29**:5078-5086. DOI: 10.1016/j.vaccine.2011.04.073

[27] Vilte DA, Larzábal M, Mayr UB, Garbaccio S, Gammella M, Rabinovitz BC, et al. A systemic vaccine based on *Escherichia coli* O157:H7 bacterial ghosts (BGs) reduces the excretion of *E. coli* O157:H7 in calves. Veterinary Immunology and Immunopathology. 2012;**146**:169-176. DOI: 10.1016/j.vetimm.2012.03.002

[28] McNeilly TN, Naylor SW, Mahajan A, Mitchell MC, McAteer S, Deane D, et al. *Escherichia coli* O157:H7 colonization in cattle following systemic and mucosal immunization with purified H7 flagellin. Infection and Immunity. 2008;**76**:2594-2602. DOI: 10.1128/IAI.01452-07

[29] Khare S, Alali W, Zhang S, Hunter D, Pugh R, Fang FC, et al. Vaccination with attenuated *Salmonella enterica* Dublin expressing *E. coli* O157:H7 outer membrane protein Intimin induces transient reduction of fecal shedding of E coli O157:H7 in cattle. BMC Veterinary Research. 2010;**6**:35. DOI: 10.1186/1746-6148-6-35

[30] Stamm I, Mohr M, Bridger PS, Schröpfer E, König M, Stoffregen WC, et al. Epithelial and mesenchymal cells in the bovine colonic mucosa differ in their responsiveness to *Escherichia coli* Shiga toxin 1. Infection and Immunity. 2008;**76**:5381-5391

[31] Moussay E, Stamm I, Taubert A, Baljer G, Menge C. *Escherichia coli* Shiga toxin 1 enhances il-4 transcripts in bovine ileal intraepithelial lymphocytes. Veterinary Immunology and Immunopathology. 2006;**113**:367-382. DOI: 10.1016/j.vetimm.2006.06.007

[32] Menge C, Wieler LH, Schlapp T, Baljer G. Shiga toxin 1 from *Escherichia coli* blocks activation and proliferation of bovine lymphocyte subpopulations in vitro. Infection and Immunity. 1999;**67**:2209-2217

[33] Schmidt N, Barth SA, Frahm J, Meyer U, Dänicke S, Geue L, et al. Decreased STEC shedding by cattle following passive and active vaccination based on recombinant *Escherichia coli* Shiga toxoids. Veterinary Research. 2018;**49**:1-15. DOI: 10.1186/s13567-018-0523-0

[34] Martorelli L, Garimano N, Fiorentino GA, Vilte DA, Garbaccio SG, Barth SA, et al. Efficacy of a recombinant intimin, EspB and Shiga toxin 2B vaccine in calves experimentally challenged with *Escherichia coli* O157:H7. Vaccine. 2018;**36**:3949-3959. DOI: 10.1016/j.vaccine.2018.05.059

[35] Isidean SD, Riddle MS, Savarino SJ, Porter CK. A systematic review of ETEC epidemiology focusing on colonization factor and toxin expression. Vaccine. 2011;**29**:6167-6178. DOI: 10.1016/j.vaccine.2011.06.084

[36] Chakraborty S, Harro C, DeNearing B, Ram M, Feller A, Cage A, et al. Characterization of mucosal immune responses to enterotoxigenic *Escherichia coli* vaccine antigens in a human challenge model: Response profiles after primary infection and homologous rechallenge with strain H10407. Clinical and Vaccine Immunology. 2016;**23**:55-64. DOI: 10.1128/CVI.00617-15

[37] Lanata CF, Fischer-Walker CL, Olascoaga AC, Torres CX, Aryee MJ, Black RE, et al. Global causes of diarrheal disease mortality in children

<5 years of age: A systematic review. PLoS One. 2013;**8**:e72788. DOI: 10.1371/journal.pone.0072788

[38] Lamberti LM, Bourgeois AL, Fischer Walker CL, Black RE, Sack D. Estimating diarrheal illness and deaths attributable to *Shigellae* and enterotoxigenic *Escherichia coli* among older children, adolescents, and adults in South Asia and Africa. PLoS Neglected Tropical Diseases. 2014;**8**:e2705. DOI: 10.1371/journal.pntd.0002705

[39] Svennerholm A-M. From cholera to enterotoxigenic *Escherichia coli* (ETEC) vaccine development. The Indian Journal of Medical Research. 2011;**133**:188-196

[40] Zhang W, Sack DA. Current progress in developing subunit vaccines against enterotoxigenic *Escherichia coli*-associated diarrhea. Clinical and Vaccine Immunology. 2015;**22**:983-991. DOI: 10.1128/CVI.00224-15

[41] Nataro JP, Kaper JB. Diarrheagenic *Escherichia coli*. Clinical Microbiology Reviews. 1998;**11**:142-201

[42] Gaastra W, Svennerholm AM. Colonization factors of human enterotoxigenic *Escherichia coli* (ETEC). Trends in Microbiology. 1996;**4**:444-452

[43] Wolf MK. Occurrence, distribution, and associations of O and H serogroups, colonization factor antigens, and toxins of enterotoxigenic *Escherichia coli*. Clinical Microbiology Reviews. 1997;**10**:569-584

[44] Qadri F, Svennerholm A-M, Faruque ASG, Sack RB. Enterotoxigenic *Escherichia coli* in developing countries: Epidemiology, microbiology, clinical features, treatment, and prevention. Clinical Microbiology Reviews. 2005;**18**:465-483. DOI: 10.1128/CMR.18.3.465-483.2005

[45] Evans DJ, Evans DG, Opekun AR, Graham DY. Immunoprotective oral whole cell vaccine for enterotoxigenic *Escherichia coli* diarrhea prepared by in situ destruction of chromosomal and plasmid DNA with colicin E2. FEMS Microbiology Immunology. 1988;**1**:9-18

[46] Evans DG, Evans DJ, Opekun AR, Graham DY. Non-replicating oral whole cell vaccine protective against enterotoxigenic *Escherichia coli* (ETEC) diarrhea: Stimulation of anti-CFA (CFA/I) and anti-enterotoxin (anti-LT) intestinal IgA and protection against challenge with ETEC belonging to heter. FEMS Microbiology Immunology. 1988;**1**:117-125

[47] Croxen MA, Law RJ, Scholz R, Keeney KM, Wlodarska M, Finlay BB. Recent advances in understanding enteric pathogenic *Escherichia coli*. Clinical Microbiology Reviews. 2013;**26**:822-880. DOI: 10.1128/CMR.00022-13

[48] Turner AK, Terry TD, Sack DA, Londono-Arcila P, Darsley MJ. Construction and characterization of genetically defined aro omp mutants of enterotoxigenic *Escherichia coli* and preliminary studies of safety and immunogenicity in humans. Infection and Immunity. 2001;**69**:4969-4979. DOI: 10.1128/IAI.69.8.4969-4979.2001

[49] Harro C, Sack D, Bourgeois AL, Walker R, DeNearing B, Feller A, et al. A combination vaccine consisting of three live attenuated enterotoxigenic *Escherichia coli* strains expressing a range of colonization factors and heat-labile toxin subunit B is well tolerated and immunogenic in a placebo-controlled double-blind phase I trial in healthy adults. Clinical and Vaccine Immunology. 2011;**18**:2118-2127. DOI: 10.1128/CVI.05342-11

[50] Darsley MJ, Chakraborty S, DeNearing B, Sack DA, Feller A, Buchwaldt C, et al. The oral, live

attenuated enterotoxigenic *Escherichia coli* vaccine ACE527 reduces the incidence and severity of diarrhea in a human challenge model of diarrheal disease. Clinical and Vaccine Immunology. 2012;**19**:1921-1931. DOI: 10.1128/CVI.00364-12

[51] Harro C, Chakraborty S, Feller A, DeNearing B, Cage A, Ram M, et al. Refinement of a human challenge model for evaluation of enterotoxigenic *Escherichia coli* vaccines. Clinical and Vaccine Immunology. 2011;**18**:1719-1727. DOI: 10.1128/CVI.05194-11

[52] Ahrén C, Jertborn M, Svennerholm AM. Intestinal immune responses to an inactivated oral enterotoxigenic *Escherichia coli* vaccine and associated immunoglobulin A responses in blood. Infection and Immunity. 1998;**66**:3311-3316

[53] Jertborn M, Ahrén C, Holmgren J, Svennerholm AM. Safety and immunogenicity of an oral inactivated enterotoxigenic *Escherichia coli* vaccine. Vaccine n.d. 1988;**16**:255-260

[54] Sack DA, Shimko J, Torres O, Bourgeois AL, Francia DS, Gustafsson B, et al. Randomised, double-blind, safety and efficacy of a killed oral vaccine for enterotoxigenic *E. coli* diarrhoea of travellers to Guatemala and Mexico. Vaccine. 2007;**25**:4392-4400. DOI: 10.1016/j.vaccine.2007.03.034

[55] Savarino SJ, Hall ER, Bassily S, Wierzba TF, Youssef FG, Peruski LF, et al. Introductory evaluation of an oral, killed whole cell enterotoxigenic *Escherichia coli* plus cholera toxin B subunit vaccine in Egyptian infants. The Pediatric Infectious Disease Journal. 2002;**21**:322-330

[56] Svennerholm A-M, Tobias J. Vaccines against enterotoxigenic *Escherichia coli*. Expert Review of Vaccines. 2008;**7**:795-804. DOI: 10.1586/14760584.7.6.795

[57] Holmgren J, Bourgeois L, Carlin N, Clements J, Gustafsson B, Lundgren A, et al. Development and preclinical evaluation of safety and immunogenicity of an oral ETEC vaccine containing inactivated *E. coli* bacteria overexpressing colonization factors CFA/I, CS3, CS5 and CS6 combined with a hybrid LT/CT B subunit antigen, administ. Vaccine. 2013;**31**:2457-2464. DOI: 10.1016/j.vaccine.2013.03.027

[58] Giuliani MM, Del Giudice G, Giannelli V, Dougan G, Douce G, Rappuoli R, et al. Mucosal adjuvanticity and immunogenicity of LTR72, a novel mutant of *Escherichia coli* heat-labile enterotoxin with partial knockout of ADP-ribosyltransferase activity. The Journal of Experimental Medicine. 1998;**187**:1123-1132. DOI: 10.1084/jem.187.7.1123

[59] Pizza M, Giuliani MM, Fontana MR, Monaci E, Douce G, Dougan G, et al. Mucosal vaccines: Non toxic derivatives of LT and CT as mucosal adjuvants. Vaccine. 2001;**19**:2534-2541

[60] Norton EB, Lawson LB, Freytag LC, Clements JD. Characterization of a mutant *Escherichia coli* heat-labile toxin, LT(R192G/L211A), as a safe and effective oral adjuvant. Clinical and Vaccine Immunology. 2011;**18**:546-551. DOI: 10.1128/CVI.00538-10

[61] Norton EB, Branco LM, Clements JD. Evaluating the A-subunit of the heat-labile toxin (LT) as an immunogen and a protective antigen against enterotoxigenic *Escherichia coli* (ETEC). PLoS One. 2015;**10**:e0136302. DOI: 10.1371/journal.pone.0136302

[62] Norton EB, Lawson LB, Mahdi Z, Freytag LC, Clements JD. The A subunit of *Escherichia coli* heat-labile enterotoxin functions as a mucosal adjuvant and promotes IgG2a, IgA, and Th17 responses to vaccine antigens. Infection and Immunity. 2012;**80**:2426-2435. DOI: 10.1128/IAI.00181-12

[63] Glenn GM, Villar CP, Flyer DC, Bourgeois AL, McKenzie R, Lavker RM, et al. Safety and immunogenicity of an enterotoxigenic *Escherichia coli* vaccine patch containing heat-labile toxin: Use of skin pretreatment to disrupt the stratum corneum. Infection and Immunity. 2007;**75**:2163-2170. DOI: 10.1128/IAI.01740-06

[64] Güereña-Burgueño F, Hall ER, Taylor DN, Cassels FJ, Scott DA, Wolf MK, et al. Safety and immunogenicity of a prototype enterotoxigenic *Escherichia coli* vaccine administered transcutaneously. Infection and Immunity. 2002;**70**:1874-1880. DOI: 10.1128/IAI.70.4.1874-1880.2002

[65] Frech SA, DuPont HL, Bourgeois AL, McKenzie R, Belkind-Gerson J, Figueroa JF, et al. Use of a patch containing heat-labile toxin from *Escherichia coli* against travellers' diarrhoea: A phase II, randomised, double-blind, placebo-controlled field trial. Lancet. 2008;**371**:2019-2025. DOI: 10.1016/S0140-6736(08)60839-9

[66] Behrens RH, Cramer JP, Jelinek T, Shaw H, von Sonnenburg F, Wilbraham D, et al. Efficacy and safety of a patch vaccine containing heat-labile toxin from *Escherichia coli* against travellers' diarrhoea: A phase 3, randomised, double-blind, placebo-controlled field trial in travellers from Europe to Mexico and Guatemala. The Lancet Infectious Diseases. 2014;**14**:197-204. DOI: 10.1016/S1473-3099(13)70297-4

[67] Riddle MS, Savarino SJ. Moving beyond a heat-labile enterotoxin-based vaccine against enterotoxigenic *Escherichia coli*. The Lancet Infectious Diseases. 2014;**14**:174-175. DOI: 10.1016/S1473-3099(13)70355-4

[68] Lundgren A, Bourgeois L, Carlin N, Clements J, Gustafsson B, Hartford M, et al. Safety and immunogenicity of an improved oral inactivated multivalent enterotoxigenic *Escherichia coli* (ETEC) vaccine administered alone and together with dmLT adjuvant in a double-blind, randomized, placebo-controlled phase I study. Vaccine. 2014;**32**:7077-7084. DOI: 10.1016/j.vaccine.2014.10.069

[69] Ruan X, Robertson DC, Nataro JP, Clements JD, Zhang W, STa Toxoid Vaccine Consortium Group. Characterization of heat-stable (STa) toxoids of enterotoxigenic *Escherichia coli* fused to double mutant heat-labile toxin peptide in inducing neutralizing anti-STa antibodies. Infection and Immunity. 2014;**82**:1823-1832. DOI: 10.1128/IAI.01394-13

[70] Purcell AW, McCluskey J, Rossjohn J. More than one reason to rethink the use of peptides in vaccine design. Nature Reviews. Drug Discovery. 2007;**6**: 404-414. DOI: 10.1038/nrd2224

[71] Savarino SJ, Poole S, Sincock SA, McVeigh A, Lee LH, Akay Y. One step beyond: A new approach for vaccines against enterotoxigenic *Escherichia coli*. In: 40th Joint Conference ofthe US-Japan Cooperative Medical Sciences Program on Cholera and Rela 2005; 2005

[72] Ruan X, Knudsen DE, Wollenberg KM, Sack DA, Zhang W. Multiepitope fusion antigen induces broadly protective antibodies that prevent adherence of *Escherichia coli* strains expressing colonization factor antigen I (CFA/I), CFA/II, and CFA/IV. Clinical and Vaccine Immunology. 2014;**21**: 243-249. DOI: 10.1128/CVI.00652-13

[73] Roy K, Hilliard GM, Hamilton DJ, Luo J, Ostmann MM, Fleckenstein JM. Enterotoxigenic *Escherichia coli* EtpA mediates adhesion between flagella and host cells. Nature. 2009;**457**:594-598. DOI: 10.1038/nature07568

[74] Roy K, Hamilton D, Allen KP, Randolph MP, Fleckenstein JM. The EtpA exoprotein of enterotoxigenic

Escherichia coli promotes intestinal colonization and is a protective antigen in an experimental model of murine infection. Infection and Immunity. 2008;**76**:2106-2112. DOI: 10.1128/IAI.01304-07

[75] Nagy B, Fekete PZ. Enterotoxigenic *Escherichia coli* in veterinary medicine. International Journal of Medical Microbiology. 2005;**295**:443-454. DOI: 10.1016/j.ijmm.2005.07.003

[76] Nguyen TD, Vo TT, Vu-Khac H. Virulence factors in *Escherichia coli* isolated from calves with diarrhea in Vietnam. Journal of Veterinary Science. 2011;**12**:159-164

[77] Nagy B, Fekete PZ. Enterotoxigenic *Escherichia coli* (ETEC) in farm animals. Veterinary Research n.d. 1999;**30**:259-284

[78] Casey TA, Nagy B, Moon HW. Pathogenicity of porcine enterotoxigenic *Escherichia coli* that do not express K88, K99, F41, or 987P adhesins. American Journal of Veterinary Research. 1992;**53**:1488-1492

[79] Wilson RA, Francis DH. Fimbriae and enterotoxins associated with *Escherichia coli* serogroups isolated from pigs with colibacillosis. American Journal of Veterinary Research. 1986;**47**:213-217

[80] Veilleux S, Dubreuil JD. Presence of *Escherichia coli* carrying the EAST1 toxin gene in farm animals. Veterinary Research. 2006;**37**:3-13. DOI: 10.1051/vetres:2005045

[81] Bertin Y, Martin C, Girardeau JP, Pohl P, Contrepois M. Association of genes encoding P fimbriae, CS31A antigen and EAST 1 toxin among CNF1-producing *Escherichia coli* strains from cattle with septicemia and diarrhea. FEMS Microbiology Letters. 1998;**162**:235-239

[82] Gyles CL. Pathogenesis of Bacterial Infections in Animals, Fourth Edition. In: Prescott JF, Songer JG, Thoen Ch O, editers. Ames, Iowa, USA: Blackwell Publishing; 2010. ISBN: 978-0-813-81237

[83] Mainil JG, Bex F, Jacquemin E, Pohl P, Couturier M, Kaeckenbeeck A. Prevalence of four enterotoxin (STaP, STaH, STb, and LT) and four adhesin subunit (K99, K88, 987P, and F41) genes among *Escherichia coli* isolates from cattle. American Journal of Veterinary Research. 1990;**51**:187-190

[84] Contrepois M, Dubourguier HC, Parodi AL, Girardeau JP, Ollier JL. Septicaemic *Escherichia coli* and experimental infection of calves. Veterinary Microbiology. 1986;**12**:109-118

[85] Deprez P, Van den Hende C, Muylle E, Oyaert W. The influence of the administration of sow's milk on the post-weaning excretion of hemolytic *E. coli* in the pig. Veterinary Research Communications. 1986;**10**:469-478

[86] Foster DM, Smith GW. Pathophysiology of diarrhea in calves. The Veterinary Clinics of North America. Food Animal Practice. 2009;**25**:13-36. DOI: 10.1016/j.cvfa.2008.10.013

[87] Moon HW, Bunn TO. Vaccines for preventing enterotoxigenic *Escherichia coli* infections in farm animals. Vaccine. 1993;**11**:213-200

[88] Wittum TW, Dewey CE. Partial budget analysis of sow *Escherichia coli* vaccination. Swine Health Production. 1996;**1**:9-13

[89] Nesta B, Pizza M. Vaccines *against Escherichia coli*. Current Topics in Microbiology and Immunology. 2018;**416**:213-242. DOI: 10.1007/82_2018_111

[90] Uehling DT, Hopkins WJ, James LJ, Balish E. Vaginal immunization of monkeys against urinary tract infection

with a multi-strain vaccine. The Journal of Urology. 1994;**151**:214-216

[91] Palaszynski S, Pinkner J, Leath S, Barren P, Auguste CG, Burlein J, et al. Systemic immunization with conserved pilus-associated adhesins protects against mucosal infections. Developments in Biological Standardization. 1998;**92**:117-122

[92] O'Hanley P, Lalonde G, Ji G. Alpha-hemolysin contributes to the pathogenicity of piliated digalactoside-binding *Escherichia coli* in the kidney: Efficacy of an alpha-hemolysin vaccine in preventing renal injury in the BALB/c mouse model of pyelonephritis. Infection and Immunity. 1991;**59**:1153-1161

[93] Cryz SJJ, Cross AS, Sadoff JC, Wegmann A, Que JU, Furer E. Safety and immunogenicity of *Escherichia coli* O18 O-specific polysaccharide (O-PS)-toxin A and O-PS-cholera toxin conjugate vaccines in humans. The Journal of Infectious Diseases. 1991;**163**:1040-1045

[94] Cross A, Artenstein A, Que J, Fredeking T, Furer E, Sadoff JC, et al. Safety and immunogenicity of a polyvalent *Escherichia coli* vaccine in human volunteers. The Journal of Infectious Diseases. 1994;**170**:834-840

[95] Russo TA, Johnson JR. Extraintestinal isolates of *Escherichia coli*: Identification and prospects for vaccine development. Expert Review of Vaccines. 2006;**5**:45-54. DOI: 10.1586/14760584.5.1.45

[96] Lorenzo-Gómez MF, Padilla-Fernández B, García-Criado FJ, Mirón-Canelo JA, Gil-Vicente A, Nieto-Huertos A, et al. Evaluation of a therapeutic vaccine for the prevention of recurrent urinary tract infections versus prophylactic treatment with antibiotics. International Urogynecology Journal. 2013;**24**:127-134. DOI: 10.1007/s00192-012-1853-5

[97] Bauer H, Darji A, Chakraborty T, Weiss S. *Salmonella*-mediated oral DNA vaccination using stabilized eukaryotic expression plasmids. Gene Therapy. 2005;**12**:364-372. DOI: 10.1038/sj.gt.3302423

[98] Uehling DT, Hopkins WJ, Elkahwaji JE, Schmidt DM, Leverson GE. Phase 2 clinical trial of a vaginal mucosal vaccine for urinary tract infections. The Journal of Urology. 2003;**170**:867-869. DOI: 10.1097/01.ju.0000075094.54767.6e

[99] Uehling DT, Hopkins WJ, Beierle LM, Kryger JV, Heisey DM. Vaginal mucosal immunization for recurrent urinary tract infection: Extended phase II clinical trial. The Journal of Infectious Diseases. 2001;**183**:S81-S83. DOI: 10.1086/318839

[100] Hopkins WJ, Elkahwaji J, Beierle LM, Leverson GE, Uehling DT. Vaginal mucosal vaccine for recurrent urinary tract infections in women: Results of a phase 2 clinical trial. The Journal of Urology. 2007;**177**:1349-1353. DOI: 10.1016/j.juro.2006.11.093

[101] Frey C, Obolensky W, Wyss H. Treatment of recurrent urinary tract infections: Efficacy of an orally administered biological response modifier. Urologia Internationalis. 1986;**41**:444-446. DOI: 10.1159/000281253

[102] Bauer HW, Alloussi S, Egger G, Blümlein H-M, Cozma G, Schulman CC, et al. A long-term, multicenter, double-blind study of an *Escherichia Coli* extract (OM-89) in female patients with recurrent urinary tract infections. European Urology. 2005;**47**:542-548. DOI: 10.1016/j.eururo.2004.12.009

[103] Bauer HW, Rahlfs VW, Lauener PA, Blebmann GS. Prevention of recurrent

urinary tract infections with immunoactive *E. coli* fractions: A meta-analysis of five placebo-controlled double-blind studies. International Journal of Antimicrobial Agents. 2002;**19**:451-456

[104] Wagenlehner FME, Ballarini S, Pilatz A, Weidner W, Lehr L, Naber KG. A randomized, double-blind, parallel-group, multicenter clinical study of *Escherichia coli* lyophilized lysate for the prophylaxis of recurrent uncomplicated urinary tract infections. Urologia Internationalis. 2015;**95**: 167-176. DOI: 10.1159/000371894

[105] Magistro G, Stief CG. Vaccine development for urinary tract infections: Where do we stand? European Urology Focus. 2018; pii: S2405-4569(18)30218-302189. DOI: 10.1016/j.euf.2018.07.034

[106] Huttner A, Gambillara V. The development and early clinical testing of the ExPEC4V conjugate vaccine against uropathogenic *Escherichia coli*. Clinical Microbiology and Infection. 2018;**24**:1046-1050. DOI: 10.1016/j.cmi.2018.05.009

[107] Mobley HLT, Alteri CJ. Development of a vaccine against *Escherichia coli* urinary tract infections. Basel, Switzerland: Pathogens; 2016;**5**:1. DOI: 10.3390/pathogens5010001

[108] Habibi M, Asadi Karam MR, Bouzari S. Evaluation of prevalence, immunogenicity and efficacy of FyuA iron receptor in uropathogenic *Escherichia coli* isolates as a vaccine target against urinary tract infection. Microbial Pathogenesis. 2017;**110**:477-483. DOI: 10.1016/j.micpath.2017.07.037

[109] Asadi Karam MR, Habibi M, Bouzari S. Use of flagellin and cholera toxin as adjuvants in intranasal vaccination of mice to enhance protective immune responses against uropathogenic *Escherichia coli* antigens. Biologicals. 2016;**44**:378-386. DOI: 10.1016/j.biologicals.2016.06.006

[110] Ghunaim H, Abu-Madi MA, Kariyawasam S. Advances in vaccination against avian pathogenic *Escherichia coli* respiratory disease: Potentials and limitations. Veterinary Microbiology. 2014;**172**:13-22. DOI: 10.1016/j.vetmic.2014.04.019

[111] Li L, Thofner I, Christensen JP, Ronco T, Pedersen K, Olsen RH. Evaluation of the efficacy of an autogenous *Escherichia coli* vaccine in broiler breeders. Avian Pathology. 2017;**46**:300-308. DOI: 10.1080/03079457.2016.1267857

[112] Han Y, Liu Q, Willias S, Liang K, Li P, Cheng A, et al. A bivalent vaccine derived from attenuated *Salmonella* expressing O-antigen polysaccharide provides protection against avian pathogenic *Escherichia coli* O1 and O2 infection. Vaccine. 2018;**36**:1038-1046. DOI: 10.1016/j.vaccine.2018.01.036

[113] Chaudhari AA, Matsuda K, Lee JH. Construction of an attenuated *Salmonella* delivery system harboring genes encoding various virulence factors of avian pathogenic *Escherichia coli* and its potential as a candidate vaccine for chicken colibacillosis. Avian Diseases. 2013;**57**:88-96. DOI: 10.1637/10277-061312-Reg.1

[114] Van Goor A, Stromberg ZR, Mellata M. A recombinant multi-antigen vaccine with broad protection potential against avian pathogenic *Escherichia coli*. PLoS One. 2017;**12**:e0183929. DOI: 10.1371/journal.pone.0183929

[115] Uotani Y, Kitahara R, Imai T, Tsutsumi N, Sasakawa C, Nagai S, et al. Efficacy of an avian colibacillosis live vaccine for layer breeder in Japan. The Journal of Veterinary Medical Science. 2017;**79**:1215-1219. DOI: 10.1292/jvms.17-0189

[116] Nagano T, Kitahara R, Nagai S. An attenuated mutant of avian pathogenic *Escherichia coli* serovar O78: A possible live vaccine strain for prevention of avian colibacillosis. Microbiology and Immunology. 2012;**56**:605-612. DOI: 10.1111/j.1348-0421.2012.00482.x

[117] La Ragione RM, Woodward MJ, Kumar M, Rodenberg J, Fan H, Wales AD, et al. Efficacy of a live attenuated *Escherichia coli* O78:K80 vaccine in chickens and turkeys. Avian Diseases. 2013;**57**:273-279. DOI: 10.1637/10326-081512-Reg.1

[118] Kwaga JK, Allan BJ, van der Hurk JV, Seida H, Potter AA. A *carAB* mutant of avian pathogenic *Escherichia coli* serogroup O2 is attenuated and effective as a live oral vaccine against colibacillosis in turkeys. Infection and Immunity. 1994;**62**:3766-3772

[119] Kariyawasam S, Wilkie BN, Gyles CL. Construction, characterization, and evaluation of the vaccine potential of three genetically defined mutants of avian pathogenic *Escherichia coli*. Avian Diseases. 2004;**48**:287-299. DOI: 10.1637/7093

[120] Gregersen RH, Christensen H, Ewers C, Bisgaard M. Impact of *Escherichia coli* vaccine on parent stock mortality, first week mortality of broilers and population diversity of *E. coli* in vaccinated flocks. Avian Pathology. 2010;**39**:287-295. DOI: 10.1080/03079457.2010.495744

Chapter 7

Photodynamic Inactivation of *Escherichia coli* with Cationic Porphyrin Sensitizers

Jin Matsumoto, Tomoko Matsumoto, Kazuya Yasuda and Masahide Yasuda

Abstract

The activity of singlet-oxygen sensitizers for photodynamic inactivation (PDI) of microorganisms and photodynamic therapy of tumor cells has been evaluated using *Escherichia coli*, *Saccharomyces cerevisiae*, and human cancer cell lines. In this chapter, drug resistance of *E. coli* was examined based on the PDI activity of a variety of RPy-P-porphyrin sensitizers with different number of ionic valence and different hydrophobic characters. The PDI activities toward *E. coli* were evaluated using the minimum effective concentrations ($[P]$) of the porphyrin sensitizers. It was found that the $[P]$ value for *E. coli* was larger than that for *S. cerevisiae*. *E. coli* has drug-resistance toward hydrophobic and mono-cationic porphyrins. However, *E. coli* has weak drug-resistance toward the porphyrins with both polycationic character and hydrophobicity. Since the outer membrane mainly consists of lipopolysaccharides and phospholipids that are negatively charged, cationic porphyrins are able to adsorb to the outer leaflet. Then the cationic porphyrins with hydrophobic character can interact with not only the outer leaflet but also inner leaflet of the outer membrane and the plasma membrane. Thus, porphyrins may be incorporated inside *E. coli* cells via the self-promoted uptake pathway. Moreover, polycationic porphyrins can interact with DNA and proteins by strong binding affinities.

Keywords: PDT sensitizer, singlet oxygen, porphyrins, PDI activity, *Escherichia coli*, *Saccharomyces cerevisiae*

1. Introduction

Singlet-oxygen ($^1O_2$) sensitizers for photodynamic inactivation (PDI) of microorganisms and photodynamic therapy of tumor cells have been developed using *Escherichia coli*, *Saccharomyces cerevisiae*, and human cancer cell lines (e.g., HeLa cell) as model cells [1–4]. As *E. coli* is a Gram-negative bacterium, the cell wall consists of an inner membrane, cytoplasmic membrane, a periplasmic space with a peptidoglycan layer, and an outer membrane [5]. Since the *E. coli* cell wall has a low permeability, there are only a few $^1O_2$-sensitizers that can permeate the cell wall and inactivate
E. coli efficiently at low concentrations.

PDI refers to the use of a visible-light source, oxidizing agents (e.g., O_2), and photosensitizers. Photosensitizers absorb light energy that causes an energy transfer

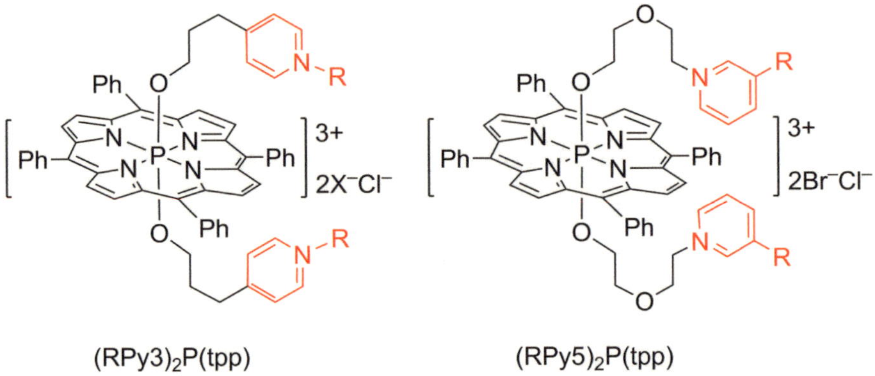

Figure 1.
Typical structure of porphyrin sensitizer (P type).

to O_2, which leads to the formation of reactive oxygen such as $^1O_2$, thereby inactivating cells and bacteria. Preliminary studies on the photodynamic action for biological systems started in the 1930s by PDI of phages using methylene blue [6, 7]. PDI of bacteria has received considerable attention as a methodology leading to the medical application of infection therapy beyond antimicrobial resistance. Among the large variety of photosensitizers developed for PDI over the last 60 years, porphyrins and metalloporphyrins became attractive sensitizers owing to their strong absorption band in the visible-light region [8–11].

In the case of porphyrin sensitizers, their solubilities in water are an important characteristic for handling them as aqueous solutions, since porphyrin derivatives, in general, are poorly soluble in water. The most popular method to improve the solubility in water is the introduction of ionic groups to the porphyrin ring. Especially, the introduction of an alkylpyridinium (RPy) group into porphyrins is a useful method to make porphyrins water-soluble [12, 13]. A typical RPy-bonded porphyrin is represented by *meso*-tetra[4-(1-methyl-pyridinium)] porphyrin (TMP). The first application of TMP to PDI was reported by Ben Amor et al. in 1998 [14]. For the last two decades, a variety of RPy-bonded porphyrins have been prepared and studied for PDI [15–21].

We have interested in axially RPy-bonded tricationic P-porphyrins and their PDI activity [22–26]. It is advantageous that the water solubilization is easily achieved through the modification of the axial ligands of P-porphyrins. It is expected that polycationic porphyrins have strong binding affinities to DNA [27–32]. In this chapter, drug resistance of *E. coli* was discussed based on PDI activity of a variety of P- and Sb-porphyrin sensitizers with different number of ionic valence and different hydrophobic character. The typical structure of the porphyrin sensitizer is shown in **Figure 1**, and they are named P-type porphyrin.

2. Materials and methods

2.1 Axially RPy-bonded tricationic P-porphyrins: $(RPy3)_2P(Tpp)^{3+}$

The preparation of tricationic bis[3-(1-alkyl-4-pyridinio)propoxo] tetraphenylporphyrinatophosphorus(V) complex, $(RPy3)_2P(Tpp)^{3+}$ (Tpp = tetraphenylporphyrinato group), was performed as follows [22]. Dichloro(tetraphenylporphyrinato)phosphorus chloride ($[Cl_2P(Tpp)]Cl$ [33], 300 mg) was reacted with 3-(4-pyridyl)-1-propanol (5.0 mL) in MeCN (30 mL) at reflux temperature for about 24 h until the Soret band shifted from 435 to 428 nm. Bis[3-(4-pyridyl)propoxo]tetraphenylporphyrinatophosphorus(V) chloride, $(Py3)_2P(Tpp)^+$, was produced in 47% yield. The $(Py3)_2P(Tpp)^+$ (50 mg) was reacted with alkyl halides (1.0 mL) in MeCN (25 mL) at reflux temperature for about 24 h to give $(RPy3)_2P(Tpp)^{3+}$ [22]. The yields of $(RPy3)_2P(Tpp)^{3+}$ are listed in **Table 1**.

2.2 Axially RPy-bonded polycationic Sb-porphyrins

Axially RPy-bonded polycationic Sb-porphyrins were prepared using dibromo(tetraphenylporphyrinato)antimony bromide ($[Br_2Sb(Tpp)]Br$) as the starting material [34]. The partial methanolysis of $[Br_2Sb(Tpp)]Br$ (1.077 g) was performed in MeOH-MeCN (1:1, 160 mL) in the presence of pyridine (0.75 mL) at 80°C until the Soret band shifted from 427 to 423 nm. Bromo(methoxo)-(tetraphenylporphyrinato)antimony bromide ([MeO(Br)Sb(Tpp)]Br, 520 mg) was formed in 61% yield [35]. An MeCN (20 mL) solution of $[Br_2Sb(Tpp)]Br$ (150 mg) and [MeO(Br)Sb(Tpp)]Br (180 mg) was heated with 3-(4-pyridyl)-1-propanol (3.7 mL) at refluxing temperature for about 24 h until the Soret band

Sensitizers	n [b]	Z [a]	Metal	Yield /%	$\varepsilon/10^4$ M^{-1} cm^{-1} [c]		C_W/mM [d]
					Soret	Q	
$(MePy3)_2P(tpp)$	1	+3	P	95	26.9	1.38	3.4
$(BuPy3)_2P(tpp)$	4	+3	P	93	23.1	1.18	6.1
$(PentPy3)_2P(tpp)$	5	+3	P	32	27.2	1.32	3.8
$(HexPy3)_2P(tpp)$	6	+3	P	47	31.3	1.45	5.8
$(HeptPy3)_2P(tpp)$	7	+3	P	32	26.7	1.26	6.0
$(OctPy3)_2P(tpp)$	8	+3	P	48	18.7	0.97	3.8
$(HexPy3)_2Sb(tpp)$	6	+3	Sb	35	16.3	4.18	11.1
(MePy3)Sb(tpp)	1	+2	Sb	42	12.7	4.45	2.4
(HexPy3)Sb(tpp)	6	+2	Sb	25	15.1	4.48	5.2
$(MePy5)_2P(tpp)$	1	+3	P	73	28.2	1.36	>120
$(EtPy5)_2P(tpp)$	2	+3	P	58	29.6	1.40	>120
$(ButPy5)_2P(tpp)$	4	+3	P	44	25.3	1.29	112
$(HexPy5)_2P(tpp)$	6	+3	P	44	24.7	1.22	64
$(4EtPy5)_2P(tpp)$	2	+3	P	72	12.7 [e]	0.57 [e]	>120
$(Me)_2P(PyHex)$	6	+2	P	57	22.6	1.31	5.0
$(Me1)_2P(PyHex)$	6	+2	P	78	14.1	0.89	11.4
$(Bu1)_2P(PyMe)$	1	+2	P	94	18.1	1.01	13.6
$(Bu2)_2P(PyMe)$	1	+2	P	32	21.7	1.21	13.0
$(Hex2)_2P(PyMe)$	1	+2	P	45	28.6	1.63	8.0

[a]*Z = charge of the complex.*
[b]*n = carbon number of the alkyl chain on the Ap.*
[c]*Molar absorption coefficient for the Soret and the Q bands in MeOH solution.*
[d]*C_W = water solubility in mM.*
[e]*Broadening of UV spectra occurred.*

Table 1.
PDI of E. coli with cationic porphyrins.

shifted to 418 nm, respectively. Thus, bis[3-(4-pyridyl)propoxo]tetraphenylporphyrinatoantimony (V) bromide ($(Py3)_2Sb(Tpp)^+$, 83 mg) and 3-(4-pyridyl) propoxo(methoxo)tetraphenylporphyrinatoantimony (V) bromide ($Py3Sb(Tpp)^+$, 90 mg) were obtained in 50% and 43% yields, respectively. $(Py3)_2Sb(Tpp)^+$ (50 mg) was reacted with 1-bromohexane (0.5 mL) in MeCN (13 mL) at reflux temperature for about 24 h to give bis[3-(1-hexyl-4-pyridinio)-1-propoxo]-5,10,15,20-tetraphenylporphyrinatoantimony (V) tribromide ($(HexPy3)_2Sb(Tpp)^{3+}$, 20 mg, 35%). The reaction of ($Py3Sb(Tpp)^+$, 50 mg) with MeI and 1-bromohexane (0.5 mL in MeCN (13 mL) at reflux temperature for about 24 h gave α-(methoxo)-β-[3(1--methyl-4-pyridinio)-1-propoxo]-5,10,15,20-tetraphenylporphyrinatoantimony (V) dibromide ($MePy3Sb(Tpp)^{2+}$, 25 mg, 42%) and α-(methoxo)-β-[3 (1-hexyl-4-pyridinio)-1-propoxo]-5,10,15,20-tetraphenyl-porphyrinatoantimony (V) dibromide ($HexPy3Sb(Tpp)^{2+}$, 20 mg, 25%), respectively [24].

2.3 Axially RPy-bonded tricationic P-porphyrins: $(RPy5)_2P(Tpp)^{3+}$

Bis[5-(3-alkyl-1-pyridinio)-3-oxapentyloxo]tetraphenylporphyrinatophosphorus(V) dibromide, chloride ($(RPy5)_2P(Tpp)^{3+}$) was prepared from dihydroxo(tetraphenylporphyrinato)phosphorus chloride ($[(HO)_2P(Tpp)]Cl$), which was prepared by hydrolysis of $[Cl_2P(Tpp)]Cl$ (300 mg) by refluxing in a mixed solvent of MeCN (160 mL) with pyridine (60 mL) and H_2O (60 mL) [22]. Alkylation of $[(HO)_2P(Tpp)]Cl$ (80 mg) with di(2-bromoethyl) ether (1 mL) was performed in the presence of K_2CO_3 (19 mg) and 18-crown-6 ether (4.2 mg) in MeCN (5 mL) at 50°C to give bis(5-bromo-3-oxa-pentyloxo)tetraphenylporphyrinatophosphorus(V) chloride ($(Br5)_2P(Tpp)^+$). The $(Br5)_2P(Tpp)^+$ (50 mg) was reacted with 3-alkylpyridine (1.0 mL) in MeCN (10 mL) under heating at 100°C for 20–68 h for the preparations of $(RPy5)_2P(Tpp)^{3+}$ [22]. Similarly, bis[5-(4-ethyl-1-pyridinio)-3-oxapentyloxo]tetraphenylporphyrinatophosphorus(V) dibromide, chloride, $(4EtPy5)_2P(Tpp)^{3+}$ was prepared via the reaction of $(Br5)_2P(Tpp)^+$ (63 mg) with 4-ethylpyridine (1.0 mL) in dry MeCN (10 mL) at 100°C for 20 h.

2.4 RPy-bonded dicationic P-porphyrins at *meso* position: $(R'm)_2P(RPyTpp)^{2+}$

At first, 5,10,15-triphenyl-20-(4-pyridinyl)porphyrin (PyTpp) was prepared by reaction of pyrrole (1.55 mL), benzaldehyde (1.83 mL), and 4-formylpyridine (0.56 mL) in propanoic acid (100 mL) in an oil bath heated at 140°C for 1 h to give PyTpp (533 mg, 14%) [24]. PyTpp (101 mg) was reacted with phosphoryl chloride ($POCl_3$, 2.0 mL) in pyridine (10 mL) in a pressure bottle heated at 180°C for 1 day to give dichloro[triphenyl(4-pyridinyl)porphyrinato]phosphorus chloride ($[Cl_2P(PyTpp)]Cl$, 99.0 mg) in 81% yield. Substitution of the axial chloro ligand with a methoxo group was performed by refluxing $[Cl_2P(PyTpp)]Cl$ (82.7 mg) in MeOH (20 mL)-pyridine (0.25 mL) for 3 days until the Soret band shifted from 435 to 424 nm. Dimethoxo[5-(1-hexyl-4-pyridinio)-10,15,20-triphenyl-porphyrinato] phosphorus (V) dichloride ($(Me)_2P(HexPyTpp)^{2+}$) was prepared by reaction of $[(MeO)_2P(PyTpp)]Cl$ (62.0 mg) with 1-iodohexane (2 mL) in DMF (5 mL) in the presence of K_2CO_3 (19 mg) at 100°C for 2 h. $(Me)_2P(HexPyTpp)^{2+}$ was purified through anion exchange with chloride ions, as follows. An aqueous solution (10 mL) of $AgBF_4$ (115 mg) was added to a MeCN-MeOH (1:1 v/v, 20 mL) solution of the porphyrins. After stirring for 24 h at room temperature, the solution was washed with water (100 mL) and an aqueous NaCl solution (100 mL) three times and subjected to precipitation with hexane (200 mL) [24].

$[Cl_2P(PyTpp)]Cl$ (78–100 mg) was reacted with ethylene glycol derivatives ($H(OCH_2CH_2)_mOR'$, R' = Me, *n*-Bu, *n*-Hex, 5–7 mL) in MeCN (10 mL) in the presence of pyridine (0.75 mL) for 24 h to give bis(2-alkyloxyethoxo)-5-(4-pyridinyl)-10,15,20-triphenylporphyrinatophosphorus (V) chloride ($[(R'm)_2P(PyTpp)]Cl$) in 66–88%. Bis(2-methoxyethoxo)-5-(1-hexyl-4-pyridinyl)-10,15,20-triphenylporphyrinatophosphorus (V) bromide, chloride ($(Me1)_2P(HexPyTpp)^{2+}$) was prepared by reaction of $[(Me1)_2P(PyTpp)]Cl$ (51 mg) with 1-iodohexane (2 mL) in DMF (5 mL) in the presence of K_2CO_3 (19 mg) in an oil bath heated at 100°C for 2 h. After anion-exchange, dichloride salt of $(Me1)_2P(HexPyTpp)^{2+}$ (27 mg, 78%) was obtained. Also, other *meso*-RPy-bonded dicationic P-porphyrins (61–90 mg) were reacted with MeI (1.2 mL) in DMF (7.5 mL) in the presence of K_2CO_3 (43 mg) by heating at 100°C for 24 h to give an *N*-methyl-substituted complex. After anion exchange, $(Me1)_2P(HexPyTpp)^{2+}$ (35 mg, 94%), $(Bu2)_2P(MePyTpp)^{2+}$ (13.7 mg, 32%), and $(Hex2)_2P(MePyTpp)^{2+}$ (28.0 mg, 45%) were formed [24].

2.5 Preparation of *E. coli* suspension

E. coli K-12 (IFO 3301) was cultured aerobically at 30°C for 8 h in a LB medium (pH 6.5) consisting of bactotryptone (10 g L^{-1}), yeast extract (5 g L^{-1}), and NaCl (10 g L^{-1}). After centrifugation of the cultured broth at 12,000 rpm for 10 min, the harvested cells were washed with physiological saline (NaCl, 7 g L^{-1}) and then suspended in physiological saline, resulting in a cell suspension of *E. coli*. The cell concentrations were determined using a calibration curve and turbidity quantified by the absorbance measured at 600 nm on an UV–Vis spectrometer [24].

2.6 PDI of *E. coli*

PDI of *E. coli* was performed as follows. A phosphate buffer (0.1 M, pH 7.6) was prepared by dissolving Na_2HPO_4 (2.469 g) and NaH_2PO_4 (0.312 g) in 100 mL of water. The suspension of *E. coli* cells (1×10^5 cells mL^{-1}, 1.0 mL), an aqueous solution of the studied sensitizers (25–100 μM, 0.1 mL), and the phosphate buffer (0.1 M, pH 7.6, 8.9 mL) were introduced into L-type glass tubes, resulting in a buffer solution (10 mL) containing *E. coli* (1×10^4 cells mL^{-1}) and the studied sensitizers (0.25–1.0 μM). Under dark conditions, the L-type glass tubes were set on a reciprocal shaker and shaken at 160 rpm at room temperature for 2 h [24]. And then the L-type glass tubes were irradiated using a fluorescent lamp (Panasonic FL-15ECW, Japan; wave length = 400–723 nm; the maximum intensity: 545 nm; 10.5 W cm^{-2}) on a reciprocal shaker at room temperature. A portion of the reaction mixture (0.1 mL) was taken up to 2 h at 20-min intervals and plated on LB plates. The LB plates were incubated for 30 h at 30°C.

The amount of the living cells (B) was defined as the average number of *E. coli* colonies that appeared after an incubation period of 30 h in three replicate plates. The B values for the PDI sensitizers were recorded at each irradiation time.

2.7 Fluorescence imaging

Incorporation of porphyrin sensitizers inside cells can be examined by fluorescence microscopy images of *E. coli* on a confocal laser scanning microscope (CLSM) under laser excitation at 543 nm. The aqueous solution containing the porphyrin sensitizers and *E. coli* was incubated for 3 h at 25°C. The concentrated solution was sandwiched between a cover slip and an agar pad on a bottom cover slip to maintain its position within the same focal plane [36].

3. Results

3.1 Properties of RPy-bonded P-porphyrins

Figure 2 shows the structures of the prepared porphyrins, which were water soluble due to cationic complexes. The water solubility (C_W) is listed in **Table 1**. In addition, **Table 1** lists the absorption coefficient (ε) of Soret band around 431 nm and Q-band at 562 nm in MeOH. These porphyrins could absorb strongly visible

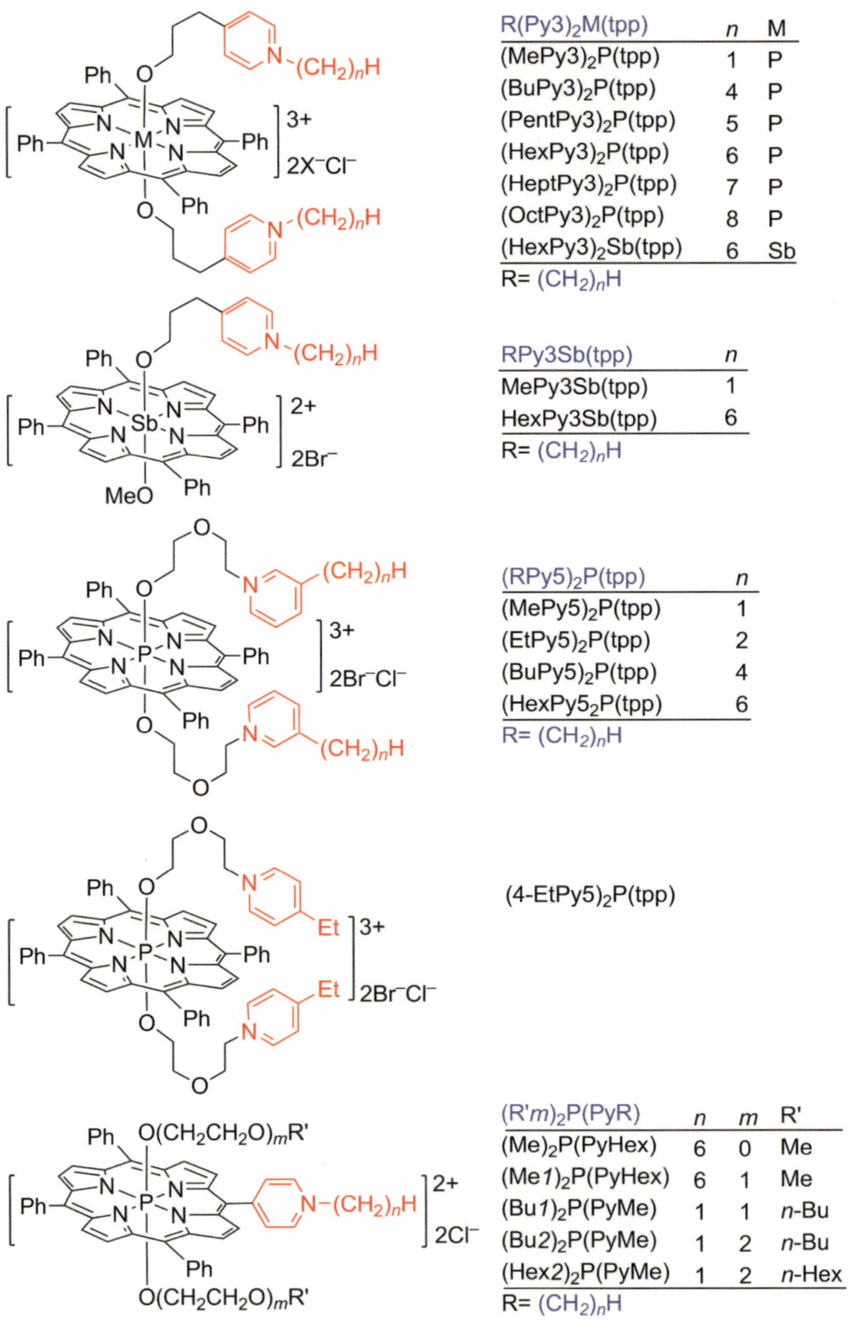

Figure 2.
Polycationic P- and Sb-porphyrins bonded to alkylpyridinium (RPy).

light. Moreover, they could generate $^1O_2$ efficiently, since the quantum yields for the formation of $^1O_2$ were found to be 0.88 for $(HexPy3)_2P(Tpp)^{3+}$ and 0.87 for $(Bu2)_2P(MePyTpp)^{2+}$ [23].

3.2 Results of PDI of *E. coli*

Results of PDI of *E. coli* are summarized in **Table 2**. As seen from **Table 2**, *Meso*-RPy-substituted P-porphyrins ($(R'm)_2P(RPyTpp)^{2+}$) have cytotoxicity, since *E. coli* was inactivated under dark conditions.

Based on **Table 2**, the survival ratios were calculated as $100B/B_0$ where B_0 is the initial amount of bacteria. From the time-course plots of survival ratios ($100B/B_0$), the half-life ($T_{1/2}$ in min), i.e., the time required to reduce B from B_0 to $0.5B_0$, was measured. A typical example of time-course plots is the case of PDI of *E. coli* by $(HexPy3)_2P(Tpp)^{3+}$ as shown in **Figure 3**. In this case, the $T_{1/2}$ value of

Sensitizers	***[P]*/ µM** [b]	**Amount of bacteria ([*B*])/CFU mL^{-1}** [a]						
		***t* = 0/min** [c]	**20**	**40**	**60**	**80**	**100**	**120**
$(MePy3)_2P(tpp)$	2.0	512 ± 22	450 ± 14	383 ± 13	344 ± 20	198 ± 13	103 ± 4.5	27 ± 1.2
$(BuPy3)_2P(tpp)$	2.0	377 ± 56	216 ± 10	105 ± 9.9	39 ± 5.3	18 ± 3.2	6.0 ± 2.7	2.3 ± 0.6
$(PentPy3)_2P(tpp)$	0.5	105 ± 12	65 ± 12	36 ± 4.6	19 ± 3.8	14 ± 4.0	11 ± 3.1	7.0 ± 2.0
$(HexPy3)_2P(tpp)$	0.5	243 ± 23	156 ± 5.2	125 ± 5.8	86 ± 3.1	77 ± 7.5	60 ± 1.2	17 ± 6.0
$(HeptPy3)_2P(tpp)$	0.4	203 ± 16	117 ± 9.1	53 ± 3.8	39 ± 3.1	15 ± 1.2	4.7 ± 2.1	3.0 ± 0
$(OctPy3)_2P(tpp)$	0.5	294 ± 14	215 ± 15	194 ± 12	136 ± 16	103 ± 9.9	76 ± 10	44 ± 8.0
$(HexPy3)_2Sb(tpp)$	1.0	152 ± 7.1	110 ± 4.7	76 ± 17	49 ± 4.2	36 ± 15	21 ± 4.5	45 ± 8.7
(MePy3)Sb(tpp)	1.0	170 ± 13	167 ± 17	134 ± 8.0	126 ± 6.8	102 ± 17	108 ± 26	113 ± 13
(HexPy3)Sb(tpp)	1.0	131 ± 28	120 ± 14	75 ± 11	55 ± 16	36 ± 11	23 ± 3.5	13 ± 1.7
$(MePy5)_2P(tpp)$	1.0	29 ± 6.4	16 ± 4.2	12 ± 5.6	10 ± 1.0	13 ± 2.3	6.7 ± 2.1	6.7 ± 1.5
$(EtPy5)_2P(tpp)$	0.25	167 ± 14	141 ± 18	59 ± 9.0	5.7 ± 0.6	1.7 ± 1.5	0.3 ± 0.6	0
$(BuPy5)_2P(tpp)$	0.25	145 ± 11	123 ± 7.6	92 ± 7.5	63 ± 4.6	33 ± 8.4	6.7 ± 4.9	4.7 ± 0.6
$(HexPy5)_2P(tpp)$	0.25	213 ± 10	213 ± 9.5	176 ± 16	166 ± 6.8	140 ± 8.2	132 ± 12	97 ± 4.4
$(4\text{-}EtPy5)_2P(tpp)$	0.5	139 ± 14	85 ± 13	88 ± 16	62 ± 6.0	42 ± 8.7	32 ± 7.0	33 ± 1.5
$(Me)_2P(PyHex)$	2.0	90 ± 13	88 ± 17	49 ± 7.8	27 ± 6.2	17 ± 5.1	13 ± 1.5	15 ± 3.1
$(Me1)_2P(PyHex)$	0.5	89 ± 2.7	57 ± 2.9	42 ± 7.2	18 ± 3.5	16 ± 2.9	8.3 ± 4.0	5.7 ± 1.2
$(Me1)_2P(PyHex)$ [d]	0.5	109 ± 26	99 ± 13	59 ± 12	64 ± 10	65 ± 165	59 ± 42	41 ± 9.6
$(Bu1)_2P(PyMe)$	0.5	24 ± 3.6	20 ± 4.5	13 ± 3.0	12 ± 1.2	7.3 ± 2.9	3.7 ± 2.1	4.7 ± 1.2
$(Bu1)_2P(PyMe)$ [d]	0.5	34 ± 5.0	25 ± 3.5	28 ± 6.1	31 ± 3.5	25 ± 1.5	20 ± 2.7	19 ± 2.1
$(Bu2)_2P(PyMe)$	2.0	126 ± 14	56 ± 3.8	21 ± 4.9	8.7 ± 2.1	3.3 ± 3.5	1.7 ± 0.6	2.3 ± 2.1
$(Bu2)_2P(PyMe)$ [d]	2.0	150 ± 13	141 ± 5.5	129 ± 8.3	124 ± 11	116 ± 13	84 ± 14	94 ± 12
$(Hex2)_2P(PyMe)$	1.0	63 ± 5.9	50 ± 7.5	56 ± 2.1	45 ± 8.1	39 ± 9.1	35 ± 6.1	33 ± 12

[a]*PDI of E. coli was performed in a phosphate buffer solution (10 mL, pH 7.6) containing E. coli (ca. 2 × 10^4 cell mL^{-1}) and porphyrin sensitizers under the irradiation of a fluorescent lamp. CFU = colony formation unit.*
[b]*[P] was adjusted to attain the value of $T_{1/2}$ between 20 and 120 min.*
[c]*Irradiation time (t) in min.*
[d]*Under dark conditions.*

Table 2.
PDI of E. coli with cationic porphyrins under visible light irradiation.

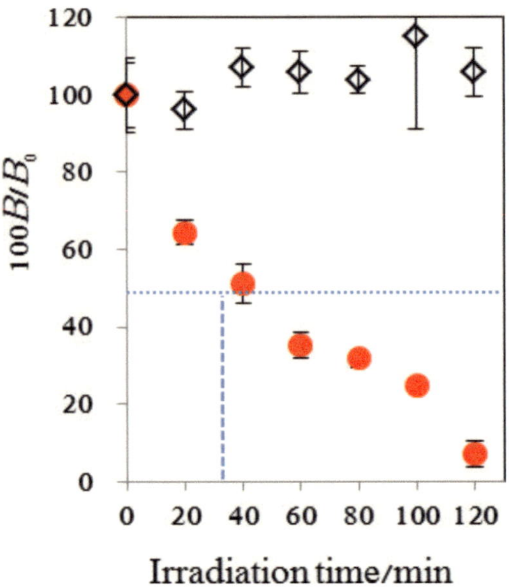

Figure 3.
Typical example of time-course plots of survival ratio ($100B/B_0$) in the PDT of E. coli with $(HexPy3)_2P(Tpp)^{3+}$ (0.5 μM) under visible light irradiation (•) and under dark conditions (◇). The $T_{1/2}$ was determined to be 31 min from the plots.

$(HexPy3)_2P(Tpp)^{3+}$ was determined to be 31 min. The minimum concentrations of the sensitizer $[P]$ were adjusted such that $T_{1/2}$ attained values between 20 and 120 min. Thus, the bactericidal activity (A_F in $\mu M^{-1}\,h^{-1}$) was evaluated using the following equation: $A_F = 60/([P] \times T_{1/2})$. **Table 3** summarizes $[P]$ and A_F values in the PDI of *E. coli*.

3.3 PDI activity of the porphyrin sensitizers toward *E. coli*

As shown in **Table 3**, the A_F values were dependent on the number of carbon atoms (n) in the alkyl group on the RPy group in $(RPy3)_2M(Tpp)^{3+}$ (M = P, Sb), $RPy3Sb(Tpp)^{2+}$, and $(RPy5)_2P(Tpp)^{3+}$. **Figure 4A** shows the dependence of the A_F values on n in the case of a series of $(RPy3)_2M(Tpp)^{3+}$ (M = P, Sb) and $RPy3Sb(Tpp)^{2+}$. The maximum value of A_F appeared at $n = 7$ whose $[P]$ value was 0.40 μM. Moderately long alkyl chain made the sensitizer more active toward *E. coli* [24]. In the case of a series of $(RPy5)_2P(Tpp)^{3+}$ (**Figure 4B**), the maximum value of A_F appeared at $n = 2$ whose $[P]$ value for *E. coli* was 0.25 μM [25]. Therefore, the A_F and $[P]$ values of 3-ethyl analog were compared with those of 4-ethyl isomer. It was found that the A_F value of 4-ethyl isomer was lower than that of 3-ethyl isomer. In the case of the 4-ethyl analog, broadening of Soret and Q bands occurred due to aggregation of porphyrin chromophores. It is suggested that aggregation caused to lower the A_F value of 4-ethyl isomer $(4EtPy5)_2P(Tpp)^{3+}$).

Figure 5 shows the fluorescence images of *E. coli* in the presence of depicting the emission from $(MePy3)_2P(Tpp)^{3+}$ and $(HexPy3)_2P(Tpp)^{3+}$ inside *E. coli*. The images show that $(HexPy3)_2P(Tpp)^{3+}$ was accumulated inside *E. coli*, whereas $(MePy3)_2P(Tpp)^{3+}$ was not. $(HexPy3)_2P(Tpp)^{3+}$, which had a large affinity to *E. coli*, had the high PDI activity. The RPy group with a long alkyl chain made the sensitizer reactive toward *E. coli*.

Sensitizer[a]	Z [b]	Metal	n [c]	$[P]/\mu M$ [d]	$T_{1/2}$/min [e]	$A_F/\mu M^{-1}\ h^{-1}$ [f]
$(MePy3)_2P(tpp)$	+3	P	1	2.0	66	0.5
$(BuPy3)_2P(tpp)$	+3	P	4	2.0	27	1.1
$(PentPy3)_2P(tpp)$	+3	P	5	0.5	29	4.1
$(HexPy3)_2P(tpp)$	+3	P	6	0.5	31	3.8
$(HeptPy3)_2P(tpp)$	+3	P	7	0.4	24	6.3
$(OctPy3)_2P(tpp)$	+3	P	8	0.5	63	1.9
$(HexPy3)_2Sb(tpp)$	+3	Sb	6	1.0	36	1.7
(MePy3)Sb(tpp)	+2	Sb	1	1.0	106	0.6
(HexPy3)Sb(tpp)	+2	Sb	6	1.0	68	0.9
$(MePy5)_2P(tpp)$	+3	P	1	1.0	40	1.5
$(EtPy5)_2P(tpp)$	+3	P	2	0.25	32	7.5
$(ButPy5)_2P(tpp)$	+3	P	4	0.25	53	4.5
$(HexPy5)_2P(tpp)$	+3	P	6	0.25	120	2.0
$(4EtPy5)_2P(tpp)$	+3	P	2	0.5	50	2.4
$(Me)_2P(PyHex)$	+2	P	6	2.0	45	0.7
$(Me1)_2P(PyHex)$	+2	P	6	0.5	37	3.2
$(Bu1)_2P(PyMe)$	+2	P	1	0.5	55	2.2
$(Bu2)_2P(PyMe)$	+2	P	1	2.0	23	1.3
$(Hex2)_2P(PyMe)$	+2	P	1	1.0	116	0.5

[a]*The PDI did not occur under dark conditions except for meso-RPy-substituted P-porphyrins, which were cytotoxic under dark conditions*
[b]*Z = charge of the complex.*
[c]*n = carbon number of the alkyl chain on the AP.*
[d]*[P] = minimum concentrations of the porphyrins adjusted to attain the value of $T_{1/2}$ between 20 and 120 min.*
[e]*$T_{1/2}$ = half-life in min.*
[f]*A_F = PDI activity in $\mu M^{-1} h^{-1}$: $A_F = 60/([P] \times T_{1/2})$.*

Table 3.
The [P], $T_{1/2}$, and A_F values in the PDI of E. coli by cationic porphyrins.

3.4 Comparison of the PDI activity in *E. coli* with the PDI activity in *Saccharomyces cerevisiae*

For comparison of the PDI activity in *E. coli* and other microorganisms, PDI of *S. cerevisiae* was performed using $(RPy3)_2P(Tpp)^{3+}$. It could photoinactivate *S. cerevisiae* in lower concentration compared with the case of *E. coli* [23]. For example, the [*P*] values of $(MePy3)_2P(Tpp)^{3+}$ for *S. cerevisiae* were 0.05 μM, while that for *E. coli* was 2.0 μM. Moreover, PDI of *S. cerevisiae* was performed using other porphyrins (Type E, **Figure 6**), which were monocationic and highly hydrophobic. The PDI of *S. cerevisiae* occurred efficiently by Type E porphyrins [37]. The [*P*] values for the PDI of *S. cerevisiae* were optimized to be 0.005 μM. Thus, *S. cerevisiae* has low drug resistance for hydrophobic sensitizers rather than polycationic sensitizers, since the [*P*] value of tricationic porphyrins was larger than that of monocationic porphyrins (Type E). On the contrary, no PDI of *E. coli* by Type E porphyrins occurred at all. This result shows that a more positive character is required for an efficient PDI of *E. coli*.

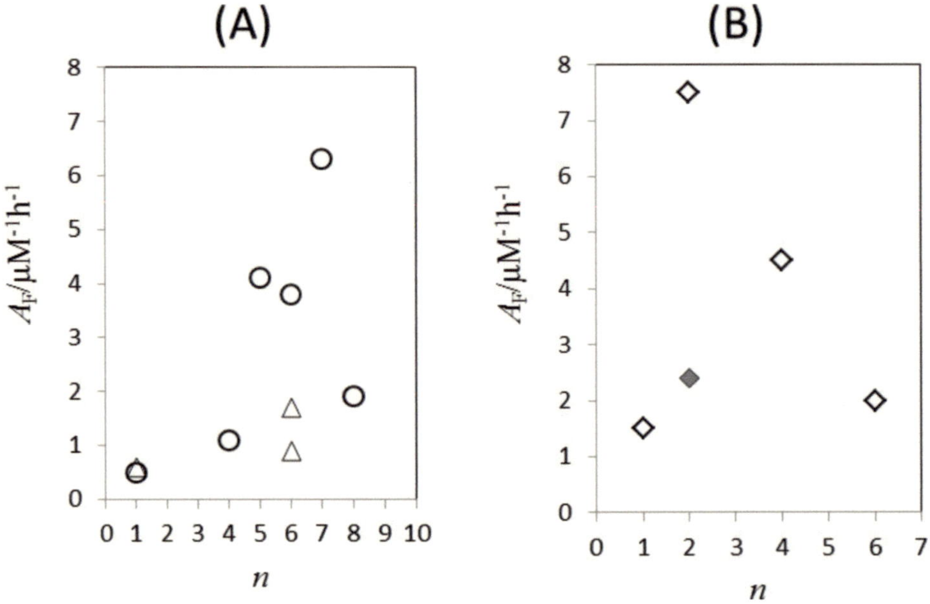

Figure 4.
Relationship between the A_F values and number of carbon atoms (n) in the alkyl group on the alkylpyridinium (RPy) in PDI of E. coli using (A) P-porphyrins ($(RPy3)_2P(Tpp)^{3+}$, ○) and Sb-porphyrins ($(RPy3)_2Sb(Tpp)^{3+}$ and $RPy3Sb(Tpp)^{2+}$, Δ) and (B) 3-alkyl-substituted P-porphyrins ($(RPy5)_2P(Tpp)^{3+}$, ◇) and their 4-ethyl-analog ($(4EtPy5)_2P(Tpp)^{3+}$, ◆).

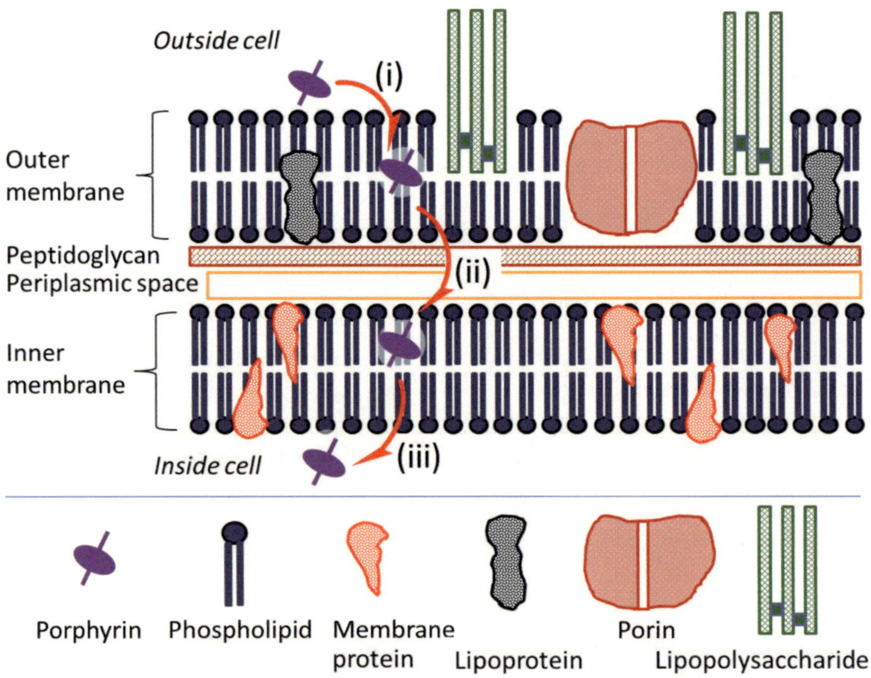

Figure 5.
The incorporation of porphyrins inside bacteria through self-promoted mechanism. (i) Cationic porphyrin adsorbs to the anionic outer membrane; (ii) amphiphilic porphyrin interacts with hydrophobic parts of outer and inner membranes; (iii) porphyrin is incorporated inside the cell.

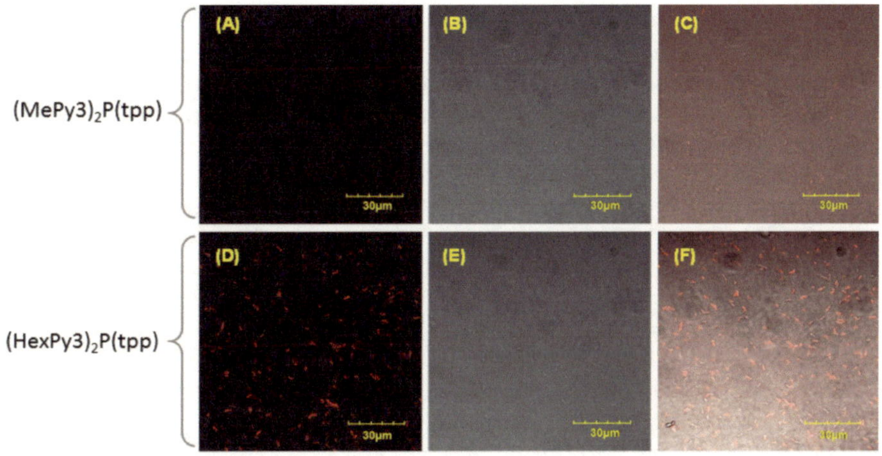

Figure 6.
Fluorescence images of E. coli obtained with a CLSM under laser-excitation at 543 nm. Fluorescence coming from inside the cells was observed with the addition of $(HexPy3)_2P(Tpp)^{3+}$ (D), but not observed with the addition of $(MePy3)_2P(Tpp)^{3+}$ (A). Transmission images of E. coli containing $(HexPy3)_2P(Tpp)^{3+}$ (E) and $(MePy3)_2P(Tpp)^{3+}$ (B). The image of C is obtained by overlapping images in A and B, and the image in F is obtained by overlapping images in D and E.

4. Discussion

The mechanism behind the PDI activity in *E. coli* is still not completely understood. However, it is known that the first contact of porphyrin photosensitizers occurs at the outer membrane. The outer leaflet of the outer membrane mainly consists of lipopolysaccharides and phospholipids, which are negatively charged and are stabilized with divalent cations such as Ca^{2+} and Mg^{2+} [38]. Therefore, electrostatic interaction between cationic photosensitizers and the outer leaflet instead of these divalent cations promotes destabilization of the outer membrane [39]. In the case of the cationic porphyrins with hydrophobic character, or the amphiphilic one, they can also interact with not only the outer leaflet but also the inner leaflet of the outer membrane and the plasma membrane (**Figure 7**). Thus, the amphiphilic porphyrins may be incorporated inside *E. coli* cells via the self-promoted uptake pathway [37]. The porphyrin

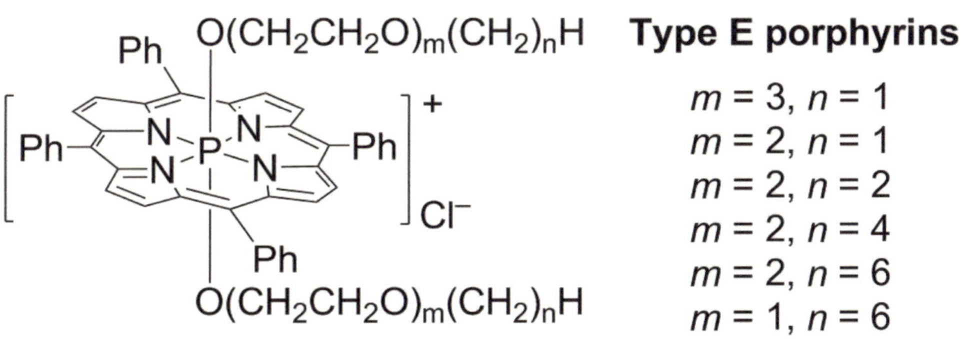

Figure 7.
P-porphyrins (Type E) substituted with alkylethyleneglycol ligands.

sensitizers passed through the cell wall may reach biogenic proteins, lipids, and DNA. Under irradiation, reactive oxygen such as $^1O_2$ was generated near to these molecules to induce cell death. Although E-type porphyrins generate $^1O_2$ efficiently under visible light irradiation, the lifetime of $^1O_2$ in aqueous medium is very short (~3 μs) [40]. Thus, for efficient PDI, $^1O_2$ should be generated as close as possible to the target molecules. The P type porphyrins with amphiphilic characters, which can be incorporated inside *E. coli*, will be advantageous to PDI via $^1O_2$ generation.

5. Conclusion

PDI of *E. coli* K-12 (IFO 3301) was examined using 19 kinds of cationic porphyrin sensitizers. In conclusion, (1) *E. coli* has high drug-resistance toward the hydrophobic and monocationic porphyrins such as Type E. (2) However, *E. coli* has low drug-resistance toward polycationic porphyrins such as Type P. (3) Especially, *E. coli* has low drug-resistance toward polycationic porphyrins with moderately long alkyl chain, for example, $(HeptPy3)_2P(Tpp)^{3+}$ and $(EtPy5)_2P(Tpp)^{3+}$. Alkyl chains might result in moderate hydrophobicity to take advantage of interaction between hydrophobic parts of cell membranes. (4) Polycationic porphyrins can interact with the anionic outer membrane at the first step and DNA and proteins inside the cells with strong binding affinities.

Acknowledgements

We thank Mr. Tomohiko Shinbara, Mr. Hiroki Kanemaru, Mr. Yusaku Suemoto, Mr. Kyosuke Takemori, Mr. Masato Shigehara, Mr. Kou Suzuki, Ms. Akari Miyamoto, and Hidekazu Uezono for their efforts on PDI of *E. coli* at University of Miyazaki.

Conflict of interest

The authors declare that they have no competing interests.

Abbreviations

A_F	PDI activity (in $\mu M^{-1} h^{-1}$): $A_F = 60/([P] \times T_{1/2})$
B	mount of bacteria
B_0	initial amount of bacteria
CFU	colony formation unit
C_W	water solubility
ε	molar absorption coefficient
LB	Luria-Bertani medium
m	number of ethylene glycol unit
n	carbon number of the alkyl chain on the Ap
$[P]$	minimum effective concentrations of sensitizer
PDI	photodynamic inactivation
RPy	*N*-alkylpyridinium group
t	irradiation time
$T_{1/2}$	half-life time required to reduce B from B_0 to $0.5B_0$
Z	valence number of the porphyrin complex

Abbreviations of substances

$(Br5)_2P(Tpp)^+$	bis(5-bromo-3-oxapentyloxo)tetraphenylporphyrinato-phosphorus chloride
$(Py3)_2P(Tpp)^+$	bis[3-(4-pyridyl)propoxo]tetraphenylporphyrinato-phosphorus chloride
$(Py3)_2Sb(Tpp)^+$	bis[3-(4-pyridyl)propoxo]tetraphenylporphyrinato-antimony bromide
$Py3Sb(Tpp)^+$	3-(4-Pyridyl)propoxo(methoxo)tetraphenylporphyrinato antimony bromide
PyTpp	triphenyl(4-pyridinyl)porphyrin
$(RPy3)_2P(Tpp)^{3+}$	bis[3-(1-alkyl-4-pyridinio)propoxo]tetraphenylporphyrinatophosphorus chloride, dihalide
$(RPy3)_2Sb(Tpp)^{3+}$	bis[3-(1-alkyl-4-pyridinio)propoxo]tetraphenylporphyrinatoantimony tribromide
$(RPy5)_2P(Tpp)^{3+}$	bis[5-(3-alkyl-1-pyridinio)-3-oxapentyloxo]tetraphenyl-porphyrinatophosphorus dibromide, chloride
$RPy3Sb(Tpp)^{2+}$	α-(methoxo)-β-[3-(1-hexyl-4-pyridinio)-1-propoxo]-5,10,15,20-tetraphenylporphyrinatoantimony (V) dibromide
$(R'm)_2P(RPyTpp)^{2+}$	bis(2-alkyloxyethoxo)-5-(1-alkyl-4-pyridinio)-10,15,20-triphenylporphyrinatophosphorus (V) dichloride
TMP	*meso*-tetra[4-(1-methylpyridinium)]porphyrin

Author details

Jin Matsumoto[1], Tomoko Matsumoto[2], Kazuya Yasuda[3] and Masahide Yasuda[1]*

1 Department of Applied Chemistry, Faculty of Engineering, University of Miyazaki, Miyazaki, Japan

2 Center for Collaborative Research and Community Cooperation, University of Miyazaki, Miyazaki, Japan

3 Department of Pharmacy, University of Miyazaki Hospital, Miyazaki, Japan

*Address all correspondence to: yasuda@cc.miyazaki-u.ac.jp

References

[1] Castano AP, Demidova TN, Hamblin MR. Mechanisms in photodynamic therapy: Part one—Photosensitizers, photochemistry and cellular localization. Photodiagnosis and Photodynamic Therapy. 2004;**1**:279-293

[2] Hamblin MR, Hasan T. Photodynamic therapy: A new antimicrobial approach to infectious disease? Photochemical and Photobiological Sciences. 2004;**3**:436-450

[3] Salmon-Divon Nitzan MY, Malik Z. Mechanistic aspects of *Escherichia coli* photodynamic inactivation by cationic tetra-*meso*(*N*-methylpyridyl)porphine. Photochemical and Photobiological Sciences. 2004;**3**:423-429

[4] Banfi S, Caruso Buccafurni EL, Battini V, Zazzaron S, Barbieri P, Orlandi V. Antibacterial activity of tetraarylporphyrin photosensitizers: An in vitro study on gram negative and gram positive bacteria. Journal of Photochemistry and Photobiology, B: Biology. 2006;**85**:28-38

[5] Alves E, Faustino MAF, Neves MGPMS, Cunha T, Nadais H, Almeida A. Potential applications of porphyrins in photodynamic inactivation beyond the medical scope. Journal of Photochemistry and Photobiology C: Photochemistry Reviews. 2014;**22**:34-57

[6] Clifton CE. Photodynamic action of certain dyes on the inactivation of Staphylococcus bacteriophage. Proceedings of the Society for Experimental Biology and Medicine. 1931;**28**:745-746

[7] Perdrau JR, Todd C. The photodynamic action of methylene blue on bacteriophage. Proceedings of the Royal Society of London Series B, Containing Papers of a Biological Character. 1933;**112**:277-287

[8] Pandey RK, Zheng G. Porphyrins as photosensitizers in photodynamic therapy. In: Kadish KM, Smith KM, Guilluy R, editors. The Porphyrin Handbook. Vol. 6. San Diego: Academic Press; 2000. pp. 157-230

[9] Nyman ES, Hynninen PH. Research advances in the use of tetrapyrrolic photosensitizers for photodynamic therapy. Journal of Photochemistry and Photobiology B: Biology. 2004;**73**:1-28

[10] Shiragami T, Matsumoto J, Inoue H, Yasuda M. Antimony porphyrin complexes as visible-light driven photocatalyst. Journal of Photochemistry and Photobiology C Photochemistry Reviews. 2005;**6**:227-248

[11] Ethirajan M, Chen Y, Joshi P, Pandey RK. The role of porphyrin chemistry in tumor imaging and photodynamic therapy. Chemical Society Reviews. 2011;**40**:340-362

[12] Kalyanasundaram K. Photochemistry of water-soluble porphyrins: Comparative study of isomeric tetrapyridyl- and tetrakis (*N*-methylpyridiniumyl)porphyrins. Inorganic Chemistry. 1984;**23**:2453-2459

[13] Girek B, Sliwa W. Porphyrins functionalized by quaternary pyridinium units. Journal of Porphyrins and Phthalocyanines. 2013;**17**:1139-1156

[14] Ben Amor T, Bortolotto L, Jori G. Porphyrins and related compounds as photoactivatable insecticides. 2. Phototoxic activity of meso-substituted porphyrins. Photochemistry and Photobiology. 1998;**68**:314-318

[15] Kano K, Fukuda K, Wakami H, Nishiyabu R, Pasternack RF. Factors influencing self-aggregation tendencies of cationic porphyrins in aqueous

solution. Journal of the American Chemical Society. 2000;**122**:7494-7502

[16] Kubát P, Lang K, Anzenbacher P Jr, Jursíková K, Král V, Ehrenberg B. Interaction of novel cationic *meso*-tetraphenylporphyrins in the ground and excited states with DNA and nucleotides. Journal of the Chemical Society, Perkin Transactions. 2000;**1**:933-941

[17] Trommel JS, Marzilli LG. Synthesis and DNA binding of novel water-soluble cationic methylcobalt porphyrins. Inorganic Chemistry. 2001;**40**:4374-4383

[18] Lang K, Mosinger J, Wagnerová DM. Photophysical properties of porphyrinoid sensitizers non-covalently bound to host molecules; models for photodynamic therapy. Coordination Chemistry Reviews. 2004;**248**:321-350

[19] Banfi S, Caruso E, Buccafurni L, Battini V, Zazzaron S, Barbieri P, et al. Antibacterial activity of tetraaryl-porphyrin photosensitizers: An in vitro study on gram negative and gram positive bacteria. Journal of Photochemistry and Photobiology B: Biology. 2006;**85**:28-38

[20] Haeubl M, Reith LM, Gruber B, Karner U, Müller N, Knör G, et al. DNA interactions and photocatalytic strand cleavage by artificial nucleases based on water-soluble gold(III) porphyrins. Journal of Biological Inorganic Chemistry. 2009;**14**:1037-1052

[21] Batinic-Haberle I, Spasojevic I, Tse HM, Tovmasyan A, Rajic Z, Clair DKS, et al. Design of Mn porphyrins for treating oxidative stress injuries and their redox-based regulation of cellular transcriptional activities. Amino Acids. 2012;**42**:95-113

[22] Matsumoto J, Kubo T, Shinbara T, Matsuda N, Shiragami T, Fujitsuka M, et al. Spectroscopic analysis of the interaction of human serum albumin with tricationic phosphorus porphyrins bearing axial pyridinio groups. Bulletin of the Chemical Society of Japan. 2013;**86**:1240-1247

[23] Matsumoto J, Kai Y, Yokoi H, Okazaki S, Yasuda M. Assistance of human serum albumin to photo-sensitized inactivation of *Saccharomyces cerevisiae* with axially pyridinio-bonded P-porphyrins. Journal of Photochemistry and Photobiology B: Biology. 2016;**161**:279-283

[24] Matsumoto J, Suemoto Y, Kanemaru H, Takemori K, Shigehara M, Miyamoto A, et al. Alkyl substituent effect on photosensitized inactivation of *Escherichia coli* by pyridinium-bonded P-porphyrins. Journal of Photochemistry and Photobiology B: Biology. 2017;**168**:124-131

[25] Matsumoto J, Yasuda M. Optimal axial alkylpyridinium-bonded tricationic P-porphyrin in photodynamic inactivation of *Escherichia coli*. Medicinal Chemistry Research. 2018;**27**:1478-1484

[26] Matsumoto J, Shiragami T, Hirakawa K, Yasuda M. Water-solubilization of P(V) and Sb(V) porphyrin and their photobiological application. International Journal of Photoenergy. 2015:148964

[27] Pasternack RF, Ewen S, Rao A, Meyer AS, Freedman MA, Collings PJ, et al. Interactions of copper(II) porphyrins with DNA. Inorganica Chimica Acta. 2001;**317**:59-71

[28] Sirish M, Chertkov VA, Schneider HJ. Porphyrin-based peptide receptors: Syntheses and NMR analysis. Chemistry—A European Journal. 2002;**8**:1181-1188

[29] Marczak R, Sgobba V, Kutner W, Gadde S, D'Souza F, Guldi DM. Langmuir-Blodgett films of a

cationic zinc porphyrin-imidazole-functionalized fullerene dyad: Formation and photoelectrochemical studies. Langmuir. 2007;**23**:1917-1923

[30] Tada-Oikawa S, Oikawa S, Hirayama J, Hirakawa K, Kawanishi S. DNA damage and apoptosis induced by photosensitization of 5,10,15,20-tetrakis (*N*-methyl-4-pyridyl)- 21H,23H-porphyrin via singlet oxygen generation. Photochemistry and Photobiology. 2009;**85**:1391-1399

[31] Kim YH, Jung SD, Lee MH, Im C, Kim YH, Jang YJ, et al. Photoinduced reduction of manganese(III) *meso*-tetrakis(1-methyl-pyridinium-4-yl) porphyrin at AT and GC base pairs. The Journal of Physical Chemistry. B. 2013;**117**:9585-9590

[32] Gyulkhandanyan A, Gyulkhandanyan L, Ghazaryan R, Fleury F, Angelini M, Gyulkhandanyan G, et al. Assessment of new cationic porphyrin binding to plasma proteins by planar microarray and spectroscopic methods. Journal of Biomolecular Structure and Dynamics. 2013;**31**:363-375

[33] Fueda Y, Suzuki H, Komiya Y, Asakura Y, Shiragami T, Matsumoto J, et al. Bactericidal effect of silica gel-supported porphyrinatophosphorus(V) catalysts on *Escherichia coli* under visible light irradiation. Bulletin of the Chemical Society of Japan. 2006;**79**:1420-1425

[34] Shiragami T, Andou Y, Hamasuna Y, Yamaguchi F, Shima K, Yasuda M. Effects of an axial ligands on reduction potentials, proton dissociation, and fluorescence quantum yield of hydroxoporphyrinatoantimony(V) complexes. Bulletin of the Chemical Society of Japan. 2002;**75**:1577-1582

[35] Shiragami T, Tanaka K, Andou Y, Tsunami S, Matsumoto J, Luo H, et al. Synthesis and spectroscopic analysis of tetraphenylporphyrinatoantimony(V) complexes linked to boron-dipyrrin chromophore on axial ligands. Journal of Photochemistry and Photobiology A: Chemistry. 2005;**170**:287-297

[36] Tanaka K, Kitamura E, Tanaka TU. Live-cell analysis of kinetochore-microtubule interaction in budding yeast. Methods. 2010;**51**:206-213

[37] Matsumoto J, Shinbara T, Tanimura S, Matsumoto T, Shiragami T, Yokoi H, et al. Water-soluble phosphorus porphyrins with high activity for visible light-assisted inactivation of *Saccharomyces cerevisiae*. Journal of Photochemistry and Photobiology A: Chemistry. 2018;**218**:178-184

[38] Wang X, Quinn PJ. Lipopolysaccharide: Biosynthetic pathway and structure modification. Progress in Lipid Research. 2010;**49**:97-107

[39] Minnock A, Vernon DI, Schofield J, Griffiths J, Parish JH, Brown SB. Mechanism of uptake of a cationic water-soluble pyridinium zinc phthalocyanine across the outer membrane of *Escherichia coli*. Antimicrobial Agents and Chemotherapy. 2000;**44**:522-527

[40] Uzdensky AB. The biophysical aspects of photodynamic therapy. Biophysics. 2016;**61**:547-557